Technology and
Economic Development

Other Titles in This Series

Credit for Small Farmers in Developing Countries, Gordon Donald

Boom Town Growth Management: A Case Study on Rock Springs–Green River, Wyoming, John S. Gilmore and Mary K. Duff

The Emergence of Classes in Algeria: Colonialism and Socio-Political Change, Marnia Lazreg

Strategies for Small Farmer Development: An Empirical Study of Rural Development in The Gambia, Ghana, Kenya, Lesotho, Nigeria, Bolivia, Colombia, Mexico, Paraguay and Peru (2 vols.), Elliot R. Morss, John K. Hatch, Donald R. Mickelwait, and Charles F. Sweet

Administering Agricultural Development in Asia: A Comparative Analysis of Four National Programs, Richard W. Gable and J. Fred Springer

Economic Development, Poverty, and Income Distribution, edited by William Loehr and John P. Powelson

The New Economics of the Less Developed Countries: Changing Perceptions in the North-South Dialogue, edited by Nake M. Kamrany

The Military and Security in the Third World: Domestic and International Impacts, edited by Sheldon W. Simon

Protein, Calories, and Development: Nutritional Variables in the Economics of Developing Nations, Bernard A. Schmitt

Food, Politics, and Agricultural Development: Case Studies in the Public Policy of Rural Modernization, edited by Raymond F. Hopkins, Donald J. Puchala, and Ross B. Talbot

A Select Bibliography on Economic Development: With Annotations, John P. Powelson

Westview Special Studies in Social, Political, and Economic Development

Technology and Economic Development: A Realistic Perspective
edited by Samuel M. Rosenblatt

The authors focus on the technological choices that confront both less developed countries (LDCs) and multinational corporations (MNCs) in establishing new industrial enterprises. They also discuss the underlying economic realities in developing countries, particularly the necessity to calculate appropriate relative factor prices for production resources, establish realistic exchange rates, and provide adequate protection of proprietary technology. The book is intended to postulate hypotheses as well as conclusions relevant to overall U.S. policy toward international flows of technology and the strategy and tactics of development.

Samuel M. Rosenblatt is project director for the International Economic Studies Institute's Technology Studies and chief economic consultant to the International Economic Policy Association. He has served on the White House Council on International Economic Policy, with the Department of Commerce's industrial analysis and regional economic development components, and as an economist with the Federal Reserve Board. The contributors include several of the United States' leading academic scholars in the fields of technology, development, and international business.

Published in cooperation with the
International Economic Studies Institute

Technology and Economic Development: A Realistic Perspective

edited by Samuel M. Rosenblatt
with a Foreword by Harlan Cleveland

Westview Press / Boulder, Colorado

This book was prepared with the support of National Science Foundation grant INT 78-08384. However, any opinions, findings, conclusions, and/or recommendations herein are those of the authors and do not necessarily reflect the views of the National Science Foundation. The copyright permission granted by the Foundation is subject to a U.S. government license.

*Westview Special Studies in
Social, Political, and Economic Development*

All rights reserved. No part of this publication may be reproduced or transmitted in any form or by any means, electronic or mechanical, including photocopy, recording, or any information storage and retrieval system, without permission in writing from the publisher.

Copyright © 1979 by International Economic Studies Institute

Published in 1979 in the United States of America by
  Westview Press, Inc.
  5500 Central Avenue
  Boulder, Colorado 80301
  Frederick A. Praeger, Publisher

Library of Congress Cataloging in Publication Data
Main entry under title:
Technology and economic development.
  (Westview special studies in social, political, and economic development)
  Includes bibliographical references.
  1. Underdeveloped areas—Technology. I. Rosenblatt, Samuel M.
T49.5.T42   338.91'72'4   79-759
ISBN 0-89158-474-9

Printed and bound in the United States of America

Contents

About the Contributors ix

Acknowledgments, *Timothy W. Stanley* xi

Foreword, *Harlan Cleveland* xiii

1. Introduction and Overview, *Samuel M. Rosenblatt* ... 1
2. Appropriate Technology: Obstacles and Opportunities, *Gustav Ranis* 23
3. Technology and Employment: Constraints on Optimal Performance, *Howard Pack* 59
4. International Transfer of Technology to Developing Countries: Implications for U.S. Policy, *Nathaniel Leff* 87
5. Technology Transfer in Practice: The Role of the Multinational Corporation, *Samuel M. Rosenblatt and Timothy W. Stanley* 109

Annex to Chapter 5: Case Studies 132

Appendixes

A. Agenda, United Nations Conference on Science and Technology for Development 173
B. Annotated Agenda, United Nations Conference on Science and Technology for Development 175

C. Agenda and Participants, International
   Economic Studies Institute Workshop,
   June 22, 1978 184

D. Senior Advisory (Steering) Committee for
   Institute Research Project on Technology
   and the World Political Economy 187

E. About the International Economic Studies
   Institute..188

F. Institute Publications 190

About the Contributors

Nathaniel H. Leff is a professor of business economics and international business at the Columbia University Graduate School of Business who has specialized in the problems of developing countries, particularly in Latin America, with a focus on Brazil and international business. His published writings include two books and more than two dozen articles. Dr. Leff has an A.B. from Harvard, a M.A. from Columbia, and a Ph.D. from MIT. He was a recipient of the Social Science Research Council award and has been an IBM fellow in international business studies as well as a visiting professor at Princeton.

Howard Pack is a professor of economics at Swarthmore College who has written widely on technological transfer and has done field work in Africa. His publications include a book and more than twenty articles and papers. Dr. Pack has a B.B.A. from CCNY and a Ph.D. from MIT. He has been an MIT and a Brookings fellow and a consultant to the World Bank, ILO, and AID. In addition to Swarthmore, Dr. Pack's teaching and research affiliations have included Yale University, the University of Pennsylvania, The Nairobi Institute of Development Studies, and Falk Institute, Jerusalem.

Gustav Ranis is a professor of economics at Yale University, where he directed the Economic Growth Center for a number of years. His work has covered all aspects of development economics, with particular regard to Asia, and involved a number of years of overseas experience. His publications include eight books and seventy articles, many of which have dealt with the role of science and technology in development.

Dr. Ranis received a B.A. from Brandeis, and a M.A. and a Ph.D. from Yale. He has been both a Social Science Research Council and a Ford Foundation fellow and served as a consultant to U.S. and international development agencies. He served as assistant administrator for policy and planning in AID from 1965 to 1967. In 1976 he organized the NAS Bicentennial Symposium Panel on "The Role of Science and Technology in Economic Development." He was recently appointed to the Council on Science and Technology for Development.

Samuel M. Rosenblatt is project director for the International Economic Studies Institute's technology studies and also chief economic consultant to the International Economic Policy Association. He is president of SMR, Inc., an economic public policy consulting firm. In his last government position, Dr. Rosenblatt was assistant director, trade and resources, of the White House Council on International Economic Policy, following seven years in the Commerce Department. He has been a professor of economics and an economist with the Federal Reserve Board. His writings include works in monetary policy, economic development, and international trade, as well as numerous contributions to official studies. Dr. Rosenblatt received his B.S. from Syracuse, and his M.A. and Ph.D. from Rutgers.

Timothy W. Stanley is president of the International Economic Studies Institute and also of the International Economic Policy Association. His professional focus has been on international political-military as well as political-economic issues; and he has served in the Office of the Secretary of Defense, on the White House staff, as defense advisor and minister at the U.S. Mission to NATO and as an arms control negotiator. He is the author or coauthor of six books and more than a dozen articles, plus numerous Institute or Association reports. Dr. Stanley received a B.A. from Yale and holds both an LL.B. and Ph.D. from Harvard. He has taught at Harvard, George Washington, and Johns Hopkins' SAIS. He is a member of the State Department Advisory Committee on Transnational Enterprises and a Director of the Atlantic Council of the United States.

Acknowledgments

This is the second major book resulting from research by the staff of the International Economic Studies Institute in its less than four years of operations. Like the *Raw Materials and Foreign Policy* project which preceded it, this is a team effort, and is a part of a larger study program which has resulted in other reports, data compilations, and *Contemporary Issues* essays.

The Institute was established at the recommendation of the Board of Directors of the International Economic Policy Association to study longer-term international economic issues of concern to Americans. It seeks to accomplish this objective with government as well as academic, foundation, and business participation and support. The institute's trustees selected technology transfer as an appropriate subject for study because of its controversial nature and its economic importance in several dimensions of U.S. international relations.

This portion of the project, dealing with the "North-South" or development dimension, was made possible by a grant from the National Science Foundation (NSF) in support of the State Department's preparations for U.S. participation in the August 1979 United Nations Conference on Science and Technology for Development. Thanks are due to the State Department and NSF officials who encouraged our revision of the institute's report to the NSF on "Obstacles and Opportunities in Technology Utilization for Development" into this book, to the editorial staff at Westview Press, to Ambassador Harlan Cleveland for his perceptive foreword, and to all who have contributed financial support to the institute's research and publications.

Particular credit for this effort belongs to Dr. Samuel M. Rosenblatt, the institute's senior economic consultant and technology project director, who also served as editor and a coauthor of this book, developing most of the case study materials and the overview. We are grateful to Professors Gustav Ranis, Howard Pack, and Nathaniel Leff, who wrote Chapters 2, 3, and 4, respectively, for their cooperation in making this an integrated and productive effort. As the institute itself takes no positions on policy issues, the views expressed in each chapter are those of the individual author. Albert P. Toner and N. Ethelyn Thompson deserve the credit, respectively, for the manuscript's editing and production supervision.

Helpful suggestions and criticisms were received from several of the institute's trustees and from members of its Senior Advisory Committee on Technology, and particularly from the experts (listed in Appendix C) who participated in the institute's June 1978 workshop. To all of the above go our thanks, and an exoneration from any responsibility for the study's conclusions or for any deficiencies in the final product, which the undersigned must accept.

Timothy W. Stanley
President
International Economic Studies Institute

Foreword

Much of the literature about "technology transfer" and "technology for development" has a disembodied quality, as though technology were detachable from the people who design it, the culture of which it is a part, the incentives that pay for it, and the people who install it and manage it and relate it to the rest of what is going on in their time and place. The purpose of this book is to bring technology for development down to earth.

Reading about technology for development is currently quite fashionable in pre-technology societies, where technology appears as a kind of magic that the rich are hoarding and not sharing with ("transferring" to) the poor. The Third World caucus—the "Group of 77" that now numbers 110 or more in United Nations assemblies—thought for a while that OPEC's oil price fixing and perhaps other commodity cartels would give them a tool to pry loose the resources for their development. But other commodities were hard to make into tools for world politics. And in practice higher oil prices not only milked the "developed" cows but siphoned from most of the "Group of 77" the foreign exchange they would have used for development; it was used instead for development and arms purchases by the oil producers. A little of the considerable surplus became aid to the non-oil-producing poor, but most of it was invested in the high-technology economies of the West.

With "resource diplomacy"—a subject explored earlier in the International Economic Studies Institute's book, *Raw Materials and Foreign Policy*—proving to be a disappointing source of leverage, technology came to be seen as a useful approach to raise in a different way the "fairness" question in international economic relations.

The writing about technology for development has come mostly from the societies which have it in profusion—societies, however, where this modern magic also has been viewed as damaging, dehumanizing, and dangerous. The humorists had already caught the mood. "Progress was a good thing once," Ogden Nash wrote, "but it went on too long." Or, as E. B. White asked in the 1920s, "Have you ever considered how complicated things can get, what with one thing always leading to another?"

The revelation that scientists bent on discovery and engineers bent over machines could make trouble, as well as progress, came as a shock. And we who live and die by technology have hastened, in our right-minded way, to tell the pre-technology folk about it. The words and music come straight out of the old spiritual: "Nobody knows the trouble I seen."

So the conditions for a collision are already in place. Planners in developing countries see technology as the key to doing something about poverty. Planners in industrial countries still think of information as a depletable resource to be conserved, and want other people's technologies to be "appropriate," or "intermediate," or something else that their own is not. The predictable collision is scheduled for the summer of 1979, at an intersection called the United Nations Conference on Science and Technology for Development.

It was under a grant from the National Science Foundation, in support of the State Department's preparations for the UN conference, that the International Economic Studies Institute engaged in an ambitious effort to examine "Obstacles and Opportunities in Technology Utilization for Development," and undertook the research on which this book is based as part of its longer-term program to look at technology's role in international trade, development, and security. The institute has had the good fortune to supplement its own expertise on international economic relations with the insights of three of the leading scholars in the field, and to draw through not-for-attribution interviews on the experience of a number of knowledgeable business executives. The findings were independently reached, but they turn out to be mutually reinforcing.

The net conclusions are pragmatic, and a breath of fresh air on a subject stale with politicized rhetoric and "blackbox" mythologies. Transnational enterprise turns out to be neither hero nor villain, neither all-important nor negligible, just a necessary if uncomfortable conveyor belt. The latest technology is not invariably inappropriate nor the most basic invariably appropriate. It is better to start from where we are, not from somewhere else.

Much can and should be done to maximize technology's contribution to development; but little will be done unless both the "northern" and "southern" sides of the transaction work with, rather than against, the underlying economic currents, and thoroughly understand the realities of international business.

Magic would be more fun. But when it comes to technology for development, economics is more effective.

Harlan Cleveland
Aspen Institute for Humanistic Studies
Princeton, New Jersey

1
Introduction and Overview

Samuel M. Rosenblatt

In August 1979 the United Nations will convene an international Conference on Science and Technology for Development (UNCSTD), which is awaited with both hope and concern. UNCSTD is one of a series of efforts to address the role of science and technology in the economic development process. While the immediate focus of the conference is on this role of science and technology, its broader context includes such fundamental matters as the changing relationships between developed and developing countries and the different perspectives from which each nation views this relationship.

Consequently, the conference must be cognizant of a host of overlapping issues. They embrace the more technical questions of science and technology enhancement in developing countries and the improved application of scientific capabilities to the needs of the developing world. They have to do with improving the network for the dissemination and utilization of the findings of science and technology within developing countries. They also deal with the contribution that technology can make to economic progress generally in the developing world, and with the institutional changes, internationally and within individual countries, that might be needed to utilize this technology more effectively for economic development purposes. The broader issues of other public policies and programs that have a direct impact on economic development, and consequently on the contribution that technology can make to this end, are also germane to the conference. In essence, while UNCSTD will be dealing directly with narrowly technical matters, in the end its success will depend on how well it relates these issues to the broader concerns of overall economic development and the varying needs of the individual developing countries.

In a sense, there is also a "hidden agenda" for the conference reflecting the controversies and confrontations of the "North-South" dialogue between the world's rich and poor countries—and especially the frustrations of the latter group over what they regard as past exploitation and monopolistic practices by the industrial countries and their international business firms. The latter feel that the developing countries tend to ignore the economic realities of the modern world and seek to establish a double standard of behavior and obligations which will retard, rather than promote, the use of modern technology for human advancement on a global basis. However, the conference comes at an interesting time, when attitudes in both camps may be evolving toward less dramatic and inflexible positions. How much this hidden agenda comes to dominate the UNCSTD proceedings may largely determine the relative success or failure of the conference.

While it once was common practice to view the separate developing countries as a homogeneous whole, it is now widely accepted that there are great differences among them in terms of their relative stages of economic and political development and their basic infrastructures, and that consequently their needs for various technologies vary considerably. Among many of the poorer countries, emphasis should continue to be placed on satisfying basic human needs, such as needs for nutrition, housing, health care, and electricity. The accomplishment of these tasks requires that the technology provided be of the simpler and more direct variety, rather than the so-called "high" technology, which is relatively new, highly research-intensive, and therefore comparatively expensive. The latter type of technology is also likely to be the most controversial from an international legal standpoint, since it is the most likely still to be under patent protection and control. It is essential that UNCSTD keep these distinctions in mind as it proceeds. (Some useful definitions of technology—what it is,

Introduction and Overview

as well as what it is not—are contained in the succeeding chapters.)

The most direct antecedent of the present conference, the United Nations Conference on the Application of Science and Technology for the Benefit of the Less Developed Areas (UNCSAT), took place in 1963. Its purposes were broadly similar to those outlined for UNCSTD. However, for a variety of reasons, this conference failed to produce the desired results. For one thing, it suffered from some overestimations of the potential contribution that technology could make to economic development. Hence, no matter what the actual outcome of the conference, persons with high and unrealistic expectations were bound to be disappointed. Another reason for the failure was rooted in the limited focus of the conference discussion. One analyst observed that UNCSAT was "a symposium presenting views on narrowly conceived scientific and technologic options. The presentations usually were isolated from the complex factors affecting the actual social settings in which 'development' takes place."[1] UNCSTD, and its member nation participants would do well to review this experience as they proceed to develop the preparations for the forthcoming conference.

Related contemporary multilateral exercises on technology which could profoundly affect the atmosphere in which UNCSTD takes place, as well as its outcome, are the efforts in the United Nations Conference on Trade and Development (UNCTAD) to draft "codes of conduct" on technology transfer and on restrictive business practices.[2] Meetings of ad hoc groups of experts have been underway since early 1973, and initially tended to be opportunities for the Group of 77, or developing country group, to engage in considerable amounts of political and ideological rhetoric and make unrealistic demands against the multinational corporations in the industrial countries. The press release issued after the conclusion of the fifth session of the Ad Hoc Group on Technology Transfer, held in July 1978, spoke to the progress made but also recognized that considerably more was needed and that resolution of differences among the Group of 77, the Group B (industrial) countries, and the Group D (Communist) countries would require "political decisions."³

Although such decisions have not yet been made, a negotiating conference nevertheless got underway in the fall of 1978 at Geneva. As this book goes to press, this conference seems unlikely to eliminate either the "square brackets" (reflecting disagreements over many of the key substantive provisions) or some of the more basic underlying differences, such as whether such codes should be voluntary or mandatory, should impose obligations on governments as well as "enterprises" (and whether the latter category includes government-owned as well as private entities), or should call for international arbitration of disputes. The disposition of this exercise in a reasonably satisfactory fashion seems a logical prerequisite for creating a conducive atmosphere for the UNCSTD discussions.

The restrictive business practices exercise has made faster progress, with considerable agreement reached among the antitrust specialists from both the developing and the industrialized countries. A final diplomatic negotiating conference is anticipated on this code within the next year and a half.

The UNCSTD agenda, as originally approved by the Economic and Social Council, contains three items (see Appendix A). The first deals with science and technology for development, the second with institutional arrangements in science and technology, and the third concerns the utilization of the United Nations and other international organizations. A much more detailed, annotated agenda was issued subsequently (see Appendix B). Its purpose was to provide guidance in the preparation of individual national papers and to help focus the ensuing discussion at the conference itself. The research presented in this study addresses a number of these items and subitems, especially the following [in 1(a)(b)]:

(i) . . . state of technological dependency; difficulties in the transfer and selection of technology; the role of multinational corporations; and (ii) . . . assessment of national measures to enhance technology transfer and adoption of integrated national policies for technology transfer and development.

Introduction and Overview

Other subitems also touched upon, in part, include:

(iii) . . . obstacles to the successful adoption of a technology transfer policy; and 1(c)(d)(ii) . . . national measures that might be taken to overcome some of the obstacles.

Appropriate Technology

It long has been recognized that the choice of technology and technical change plays a vital role in contributing to the economic development of all nations. Other key elements of such development include increases in the supply of resources and improved efficiency in their use, including technologically induced shifts in the sectoral and industrial composition of production. Institutional changes that support competitive conditions and economic incentives and objectives, as well as more traditional non-economic goals, such as educational systems, class status, and societal stability, also play a part in this process.

In the immediate aftermath of World War II, a feeling of euphoria about the world's ability to solve its economic development problems seemed to have taken hold.[4] This attitude may, in part, have been due to the success of the U.S. program of financial aid and assistance to Western Europe which, however, started the post-war reconstruction from a much higher and more sophisticated base than applied to developing countries. It also reflected an over-optimistic assessment of the role that technology could play in the developing world. With the passage of time and the accumulation of experience, disillusionment set in regarding the ease with which traditional societies could be turned around to become industrialized, modern states through the judicious use of private investment, foreign aid, and what has been termed "sweat equity," or the involvement of local human resources. Attitudes toward technology also altered. Instead of focusing upon the transference of large-scale industrial modes of production from developed to developing countries as quick and effective solutions to economic development problems, discussions turned to the use of a more appropriate type of technology. As noted by Nicolas

Jéquier, this shift had both immediate and historical derivations. "The most conspicuous of these immediate origins," he notes, "is the realisation, shared by aid-giving and aid-receiving countries alike, that development aid and a Western style of industrialization have neither fulfilled the initial hopes which were placed in them nor been fully capable of solving the basic problems of development."[5] The quest then turned to other and more appropriate forms of technology.

The United States has been among those in the forefront of this movement. Public Law 94-161, the International Development and Food Assistance Act of 1975, specified that a portion of the funds authorized for fiscal years 1976, 1977, and 1978 could be used to develop activities in the field of intermediate technology. This technology is defined as being "neither so primitive that it offers no escape from low production and low income nor so highly sophisticated that it is out of reach for poor people, and ultimately uneconomic for poor countries."[6] This attitude was more recently expressed and reconfirmed as part of the U.S. preparations for UNCSTD. Section 507(a) of the Foreign Relations Authorization Act, Fiscal Year 1978, P.L. 95-105, says:

> The President shall take appropriate steps to ensure that, at all stages of the United Nations Conference on Science and Technology for Development, representatives of the United States place important emphasis, in both official statements and informal discussions, on the development and use of light capital technologies in agriculture, in industry, and in the production and conservation of energy.[7]

The following section of this law goes on to define light capital technologies and lays out some of the intended goals and objectives that might be achieved by its expanded use.

> The term "light capital technologies" means those means of production which economize on capital wherever capital is scarce and expensive and labor abundant and cheap, the purposes being to insure that the increasingly scarce capital in the world can be stretched to help all, rather than a small minority, of the world's poor; that workers will not be displaced by sophisticated labor-saving devices where there is already

much unemployment; and further, that poor nations can be encouraged essentially to produce their own capital from surplus labor time, thus enhancing their chances of developing independently of outside help.[8]

On its face this appears to place a heavy burden on technology, in whatever form and however defined, in a manner similar to the earlier highly optimistic expectations for modern technology transfer. Indeed, the concept of "light" technology, "intermediate" technology, or "appropriate" technology, however it is labeled, seems to have evoked ideological and missionary zeal from some of its proponents. Over-selling a program in this manner may once again dash expectations and undercut a reasonable assessment of the role technology can actually play in assisting the developing world to achieve greater economic development. In this regard it should be noted that "appropriate technology" may, in some contexts, mean capital-intensive techniques and that "light" technology may embody highly modern and efficient production processes. If these distinctions are not recognized and the temptation is not avoided to prescribe doctrinaire solutions, such support for this concept as is growing in the developing countries could well be dissipated. Many of these countries have moved from negative or skeptical attitudes about appropriate technology to a cautious acceptance of its possibilities. Misspecifying or overselling it can once again cause exaggerated expectations to be followed inevitably by disappointment and cynicism.

The concept of appropriate technology has been variously defined. One definition describes it as "the set of techniques which makes optimum use of available resources in a given environment. For each process or project, it is the technology which maximizes social welfare if factors and products are shadow priced."[9] Another definition calls technology appropriate when "factor proportions . . . are roughly in line with the overall factor availabilities in an economy. The poorer the LDC the less capital (physical and human) relative to labor and, hence, the more labor intensive the 'appropriate' factor proportions would be."[10]

The second definition emphasizes the labor-intensive characteristics of appropriate technology, an approach that is taken by many advocates of appropriate technology. One limitation of both definitions is their static nature, especially when viewed from the vantage of policy analysis and policy development.[11] This point is discussed by Frances Stewart in her book, *Technology and Underdevelopment,* where she emphasizes that there is not a single alternative or appropriate technology for a developing country.[12] Rather, what is appropriate for that country must be compared with the existing techniques in use in the capital-intensive, or modern, sector as well as in the more traditional labor-intensive sectors. The simultaneous presence of both of these technologies in developing countries is at once a cause and effect of the unbalanced nature of their economies and the problems associated with these imbalances. These problems are the familiar ones of developing countries today: a dualistic economy with modern and traditional sectors; a dichotomy between urban and rural areas associated with this dualism; widespread underemployment in rural areas and growing unemployment in urban areas; generally low labor productivity; extremes in the distribution of income; and limited internal purchasing power for broadly based consumer goods with emphasis on the availability of high quality consumer goods to satisfy the small, but wealthy, upper income groups.

The approach toward a solution to these problems does not lie solely with technology in any of its forms, appropriate or otherwise, even though there is no doubt that more effort could go into making the "right" technology choice and that the range of these choices is wider than traditionally thought. This last point is developed below, and in the sections by Pack and Ranis, as are suggestions on how to improve the frequency of choice and the likelihood of success for these choices. However, these discussions should be fitted into the context into which Stewart placed them. She noted that, "the question of whether or not an 'efficient' alternative technology exists is closely tied up with the whole strategy of [a country's] development, and can only be assessed within the context of a particular strategy."[13]

Alternative techniques exist or can be readily devised; for example, if a strategy calls for a rural, small-scale, self-reliant pattern of development, then techniques can be developed. Or if the objective is to pursue a policy of rapid, export-oriented, and internationally competitive industrialization, a more advanced technology can be applied. However, each strategy, once established, tends to be self-reinforcing, and its continuation tends to make the alternative appear to be less attractive and less feasible. Stewart sums up the ultimate underlying implication of this discussion as follows:

> Implicit behind some of this discussion is what might be termed the political economy of technical choice. The political economy of a system may be defined as the distribution of the control over resources—both consumption and investment—to which it gives rise. Associated with each technique is a particular distribution of benefits.[14]

The inference from these statements is quite clear. In a very narrow and technical sense, it would appear possible for the developing countries to choose more appropriate technologies and techniques from among the given set of available techniques, and perhaps even broaden the range from which choices are made. This would have the effect of increasing the demand for labor, raising the real wage level, and increasing the volume and scope of producer and consumer goods available to satisfy the mass market. However, before a significant breakthrough and restructuring of a country's technological base could occur, decisions and actions of a non-technology nature would have to be made and taken that would redirect the energies and resources of a developing country toward a series of economic development goals and objectives that were conducive to the widespread adoption of such technologies and techniques.[15]

It should also be noted that development goals themselves are not static, although they are often so treated for analytical purposes in the literature. Rather, they are inevitably dynamic functions of the country's political processes, and may change over time in their scope and content and in their relative priorities. This argues for a country to maintain some degree of flexibility in its technological inputs.

Technology Choice—Actual and Potential

Within the broader policy constraints just described, a major issue is the extent to which a range of technology options actually exists from which the developing countries can make choices, the degree to which they exercise this prerogative, and the actual outcomes of such choices. Initially, there was a fairly widespread presumption that because of the obvious dependence of the developing countries on the advanced industrial countries for technology, the receiving countries had very little choice regarding the type of technology they could acquire. In effect, they were thought to exist in a state of "technological determinism" and "technological dependence" on the industrial countries. Therefore, the technology the developing countries were likely to obtain, especially from multinational corporations (MNCs), would be highly capital-intensive, geared to large-scale operations, and capable of turning out only high quality, sophisticated products that were more suited to the economies of the industrial countries themselves. Such technology, according to the "dependencia" literature of recent years, would control and dictate the path of economic development that the developing countries could pursue. With the passage of time, the acquisition of experience, and the accumulation of analytic studies, this attitude has changed, although doubts—and doubters—still exist.

A 1974 study by Walter Chudson and Louis Wells, prepared for the United Nations Secretariat, found some intermediate technology in selected industries in developing countries used side by side with more automated techniques.[16] They noted, however, that "most of the evidence concerns light manufacturing in which adjustment is presumably easier, and even here it occurs mainly in the 'peripheral' operations (materials storage, handling, and packaging) rather than in 'core' processes involving the physical transformation of material."[17] However, their conclusions on this point are cautious because

Introduction and Overview

of the limited research on the choice of technology by the industrialist.

More recent research has come to somewhat stronger conclusions. In a series of case studies undertaken for the International Labour Office, A. S. Bhalla concluded that, "The studies demonstrate quite clearly that substitution possibilities exist in industry in both core and ancillary operations. This conclusion, based on empirical evidence, is important, since it has often been assumed that there is no choice of techniques in manufacturing industry."[18] In a review article on the existence and application of alternative appropriate technologies in manufacturing industries in developing countries, Lawrence White also concluded that these countries had such alternatives and did not seem to be limited to contemporary capital-intensive methods of advanced industrial countries.[19] Gustav Ranis and Howard Pack, in their respective chapters of this book, also support this conclusion.[20] It should be noted that a somewhat different view persists regarding the existence of this range of choice of technologies, and especially regarding the possibilities of effectively utilizing appropriate technology for development purposes.[21]

Nevertheless, on balance, the developing countries do indeed appear to have a wide range of alternative technologies from which to choose. In effect, there is a supply of alternative technologies. These cannot necessarily be taken "as is" off some "international shelf." They may require minor modifications and the application of local ingenuity, as well as certain management techniques, but they do exist. How actively the developing countries are exercising these choices and making these adaptations is discussed below.

Demand Limitations on Choice

Within this range of technological alternatives, the question is what factors impede or encourage the adoption of appropriate techniques. As Ranis puts it: Given this wide choice, why is it that "we are nevertheless confronted with the empirical fact that the selections actually made in most of the developing countries still appear to be substantially 'inappropriate' by any known quantitative or judgmental standard."[22] Most authorities who attempt to answer this question would cite inappropriate factor prices and the absence of competitive markets in the developing countries among the major explanations for this behavior.

There are indeed differences between the capital-labor price ratios in developed and developing countries across comparable industries and goods; that is, capital is relatively more expensive in the developing countries than in the industrial countries, and labor is relatively cheaper. However, these differences are less than might be expected given the larger differences in relative factor supplies in these two groups of countries. These results come about because the cost of capital in developing countries tends to be underpriced, relative to its true or market opportunity costs, for such reasons as government subsidies, overvalued and other exchange rate policies that encourage capital imports, and tax preferences and depreciation policies. At the same time the cost of labor is inflated (again relative to its opportunity cost) by such policies as minimum wage legislation, mandated social welfare programs, and restrictions on firing employees.[23] The absence of competitive market pressure minimizes, or eliminates, the necessity for entrepreneurs, whether local or associated with a foreign multinational corporation, to seek out and adopt either the most efficient techniques by which to produce a product or the most appropriate goods needed to satisfy market demand.

There are, of course, other factors that affect the demand for appropriate technology. Among these is the market demand for particular products. This demand, in turn, is affected by the developing countries' macroeconomic policies, including their policies regarding income distribution. Such policies have a direct influence on the effective demand for various types of appropriate or inappropriate goods.[24] Other factors that limit the demand for appropriate techniques include an engineering bias toward capital intensity—a desire for the newest and best; management bias toward capital intensity— a desire for better control over sudden changes in output, given the developing country environment in which the firm operates; and product quality bias—which places emphasis on automated and controlled production techniques needed to establish and maintain high quality.[25] Finally, the very inefficient information network that exists in most developing countries leaves many entrepreneurs totally unaware of the existence of alternative, and more labor-intensive, techniques.[26] Conversely, if the costs of obtaining information are too high, it may be rational for the entrepreneur not to expend his time and energy seeking alternatives. On this point, Pack notes that multinational corporations may have an advantage, since their costs for information searches would be lower than those of locally-owned firms.[27] The multinational corporations could therefore afford to seek out a more appropriate alternative.

The scale of operation also enters heavily into this discussion of demand. On this point, Pack contends that a larger scale of output per firm need not necessarily lead to greater capital intensiveness.[28] On the other hand, Stewart has argued that scale of operation is decisive in the choice of appropriate technology and that, given the size of the market in the developing countries, small scale is often more appropriate.[29] Finally, in his chapter of the book, Nathaniel Leff contends that, regardless of the type of technology considered, the aggregate effective demand by the developing countries for this technology is likely to be quite limited.[30]

A new factor has entered the equation since OPEC's successful cartelization of the world oil supply and the quintupling of its price. The several-fold increase in the cost of imported oil has created additional constraints on the already scarce foreign exchange resources of many non-oil-producing developing countries. This has meant less money for most of the developing world to spend on imported technology of any type (which may account, in part, for the renewed emphasis in UNCTAD and elsewhere on improving the developing countries' "terms of trade" for such technology). The premium price for energy inputs has also altered the market demand for technology. Energy-intensive technologies now strain a scarce and expensive resource, and for countries like Brazil (which devotes about a third of its foreign exchange to imported oil) also involve balance of payments constraints. Energy-efficient technologies, on the other hand, are in growing demand. This change has, in a sense, enhanced the role of multinational corporations in the selection and choice process because they are the source of many of the key technologies for both energy production and its industrial utilization. Their opportunities—and responsibilities—may therefore be proportionately greater.

Enhancing Technology Choice Outcomes

Consideration could be given to a wide variety of means of improving the supply and demand sides of the technology equation. Introducing the correct pricing signals for capital and labor and intensifying the degree of competitive market pressures would undoubtedly result in more rational decision making within the developing countries. Similarly, creating a more effective aggregate demand for a wider variety of appropriate goods and services that are compatible with the social, economic, energy, and natural environments in which the developing countries operate would also be conducive to this end. Decisions on these options must be made principally by the developing countries themselves. U.S. and other industrial countries' policies could be more supportive of these choices, however, taking into account the learning curves that are inevitably involved.

Vast impediments exist in the developing countries to a proper flow of information between the developers and users of technology. Virtually all analysts of the problem agree on this point.[31] The research communities in these countries tend to be isolated from the potential users of their output, and, equally important, they tend to respond to their own systems of rewards that seek approval and affirmation in research communities located in the more advanced industrial countries.[32] This means they have little or no appreciation for the scientific and technological needs of their own industries. These research communities should direct more effort toward the practical needs of their domestic industries (including their energy situations) in order to encourage the local adaptation

Introduction and Overview

of imported technologies. Chudson and Wells cite a study of innovation in Argentina that illustrates the importance of these local innovations and incremental adaptations.[33] Similarly, Ranis stresses the importance of gradual and unspectacular change. Moreover, such an approach could expand the innovative and productive contribution of the science and technology community of the developing countries.[34]

Denis Goulet speaks of the so-called "Sabato triangle," named after an Argentine physicist, Jorge Sabato, that could provide a model for technology policy in this regard. This model "aims at creating practical linkages among research, production, and development-policy actors. ... Each ... must be linked by a flow of information with the other two; each must take initiatives in demanding or supplying technology."[35] According to Goulet, this concept enjoys widespread acceptance among Latin American specialists in technology.[36] The political, economic, and organizational implications of such a model are complex, however.

Technology Transfer

There are many ways to transfer technology to developing countries. One way is through direct investment by a multinational corporation in a wholly-owned or majority-owned subsidiary. Another is through a joint venture where the multinational has a minority interest. Other ways include the use of licenses and patents, turn-key operations, management contracts, equipment suppliers, and consultative arrangements that may be done independently or in combination with some of the other modes of transfer. The essence of all these methods however, is that the technology moves from a private entity in the industrial country to the recipient, public or private, in the developing country. While the developing countries may resent this dependence, "the harsh truth is that poor countries do need technology, and there exist few alternative sources outside the TNCs [transnational corporations] where they may obtain it."[37] (Such technology as may be public property, by virtue of government-financed research or otherwise, could, of course be contributed as part of government foreign assistance. But its relevance to development needs tends to be limited.)

As described below by Rosenblatt and Stanley, multinational corporations have a variety of reasons for entering into technology transfer arrangements. These include export potential, market protection, market penetration, and increased production to reduce unit costs, but the common denominator is the objective of a satisfactory economic return. Developing countries are attempting to alter their relationships with the multinational corporations, focusing on what they call the restrictive business practices of these entities, as in the aforementioned UNCTAD discussions on the transfer of technology. These discussions concentrate on such matters as patents, licenses, marketing restrictions, and other devices that, from the perspective of the developing countries, may appear to limit their economic options and slow down their economic development. The evidence on these points is mixed, but some would agree that, in the past, patents have been used in part to protect markets and restrict entry by the developing countries.[38] However, regardless of this history, the balance between the multinational corporation and the developing country has now shifted so that the developing countries are in a much less dependent position.

There is now considerably more competition among the multinational corporations in the quest for markets and entry into the developing countries. Moreover, developing countries such as the Andean Group in South America and the Association of Southeast Asian Nations (ASEAN) provide individual countries with greater bargaining power. The United Nations Center on Transnational Enterprises has as one of its purposes the providing of expert advice to developing countries in their negotiations for technology and other arrangements with international firms.

Finally, it is important to note that an active involvement by a developing country firm with a multinational corporation involves much more than the use of patented, proprietary knowledge. The multinational corporation's intangible managerial insights and "know-how" in the end may prove to be the most valuable contribution it makes to the developing country.

In the vast majority of cases, the multinational company makes its decisions in the context of a long-term relationship with the host country and particular partnership arrangements within it. Where the essential ingredients are lacking, there is little that industrial country governments can do to change the judgmental factors influencing the private sector. They can, of course, assign a higher priority to technological inputs in their public sector aid programs; but there are practical limits here as well.

Policy Framework and Recommendations

The chapters which follow develop the points discussed above. Each of the authors has approached the subject from a somewhat different vantage point and reached somewhat different conclusions. Nevertheless, throughout these papers and at the workshop that was held to review them, a certain consensus emerged on the proper approach that the United States, the other industrial countries, and the developing countries might take toward UNCSTD.[39] In many respects, then, the independent work of the several authors has reached quite parallel conclusions.

In abbreviated fashion, these are the principal elements of this consensus:

1. There should be a wariness of exaggerated claims and expectations regarding the contributions that advanced country science and technology can make toward solving the economic development problems of developing countries.
2. While much of the technology that will be used by the developing countries will come initially from the industrial countries, and specifically the multinational corporations, much of the successful adaptation of this technology will depend on the efforts of the developing countries themselves.
3. Multinational corporations, by and large, have not operated in a manner destructive of the efforts of developing countries to achieve economic development, and they have not attempted to impose a "technological

determinism" on these countries. Indeed, the evidence suggests that the multinational corporations have been at least as adaptive as indigenous firms, if not more so, in seeking to install appropriate labor-intensive and, more recently, energy-efficient, technologies.

4. The successful development of an appropriate technology policy by the developing countries will generally require a major realignment of their other economic and social policies. Responsibility for bringing this about must necessarily lie with the developing countries themselves.

5. There should be an awareness that appropriate technology is best applied at the working or operating level of a plant and that it achieves its results in an incremental and unspectacular fashion. This will require the developing countries to display considerable forbearance in their expectations of spectacular technological breakthroughs.

6. There are major differences among developing countries regarding their degree of development, their needs for technology, and their internal science and technology competence. The policies advocated by the United States and the other industrial countries must reflect these differences.

7. A major effort needs to be directed toward improving the flow of information among researchers, producers, and policymakers within developing countries so as to emphasize the availability, value, and utility of appropriate technology. Research centers of excellence should be carefully planned to focus on the practical application of technology in the production process of a limited number of industries. But this emphasis need not exclude a more imaginative search for technologies to solve particular problems important to the mass of people in a given region.

8. Given the information gaps and discontinuities noted above, industrialized country and multilateral aid programs can play a more productive role in the application of technology to development. Many industrial firms

evolve and apply technology exclusively or mainly in a direct relationship to their own products or processes and are normally willing to "transfer" it only as part of a larger and longer-term pattern of business relationships with a host country (or more specifically, with a local partner in the country). But there are some areas where the technology itself is the product, and credits for its acquisition by aid recipients—or via competitive bids for purchase by an aid agency for onward transfer to countries or to regional centers—could play an expanded role. Also, multinational corporations may be able to contribute appropriate technology innovations and expertise to development tasks as a voluntary expansion of their normal business operations, if suitable incentives can be structured.

Other major points and recommendations put forth by the individual authors include the following items:

1. Achieving a better balance between urban and rural development programs in developing countries would greatly assist in stimulating demand for appropriate processes and goods. Developed countries should take every opportunity to assist this process when their advice is sought. Energy conservation may be an especially promising field.
2. Developed countries should attempt to do a better job of assuring that their foreign aid and foreign investment support programs are administered so as to encourage the wider application of appropriate techniques in developing countries.
3. Developed countries might restructure their own tax and incentive systems in order to encourage their science and technology communities to undertake more effort to apply appropriate technology in developing countries.
4. Developing countries should reexamine their entire institutional structures for research and development, including their reward mechanisms, to assure that they

are as responsive as possible to the practical needs of their own producing community.
5. It would appear to be a misallocation of limited resources to attempt to establish a computerized technology-information system, highly centralized and bureaucratically controlled, as a means of informing developing countries of the range of technology choices they have available.
6. A more fruitful partnership between industrialized, advanced developing, and less developed countries could be created to utilize the middle group's experience and insights on technology for the benefit of the poorer countries.
7. More imaginative use of the experiences and contacts of multinational corporations with the developing countries should be encouraged, emphasizing the tangible and intangible strengths these corporations have displayed in such relationships. Heretofore, the political emphasis has been mostly on reducing the negative aspects. Perhaps not enough effort has been devoted to the possibilities of "optimizing" the affirmative contributions, which could be made either on a voluntary basis to improve investor-host country relations, or be stimulated by tax or other incentives.

The bottom line of the overall study is a truism with which UNCSTD participants must come to grips: Economic forces themselves are the best "conductor" of technology for development. Other factors can create impedances and short circuits, and these should be removed or minimized wherever possible; but neither goodwill, incentives, nor political decisions can substitute for the basic disciplines of competitive economic forces in the developing countries themselves.

Notes

1. See Mary M. Allen, "United Nations Conferences on Science and Technology for Development, 1963-1979," draft, George Wash-

Introduction and Overview 21

ington University, Washington, D.C., May 1977.

2. Ibid., p. 5.

3. UNCTAD, press release, TAD/INF/977, July 26, 1978.

4. Richard S. Eckaus, *Appropriate Technologies for Developing Countries* (Washington, D.C.: National Academy of Sciences, 1977), p. 6.

5. Nicolas Jéquier, ed., *Appropriate Technology, Problems and Promises*, Development Centre Studies (Paris: OECD, 1976), p. 25.

6. U.S. Congress, House Committee on International Relations, Hearings, Agency for International Development, *Proposal for a Program in Appropriate Technology*, 95th Cong., 1st Sess., February 7, 1977, p. 2.

7. U.S. Foreign Relations Authorization Act, Fiscal Year 1978, Public Law 95-105 (August 17, 1977).

8. Ibid.

9. David Morawetz, "Employment Implications of Industrialization in Developing Countries: A Survey," *Economic Journal*, Vol. 84 (September 1974), p. 517.

10. Lawrence J. White, "Appropriate Factor Proportions for Manufacturing in Less Developed Countries: A Survey of the Evidence," in AID (see note 6), p. 117.

11. Frances Stewart, *Technology and Underdevelopment* (Boulder, Colorado: Westview Press, 1977).

12. Ibid.

13. Ibid., p. 109.

14. Ibid., p. 110.

15. On this point note the following statement by Richard Eckaus: "Since the use of any particular technology is not an end in itself, the criteria of appropriateness for the choice of technology must be found in the goals of development. These goals are concerned not only with the volumes of output and income generated by an economy but also with the way they are produced and distributed among the population; they include, as well, particular patterns of national political change and national independence." Eckaus, *Appropriate Technologies*, p. 10.

16. Walter A. Chudson and Louis T. Wells, Jr., "The Acquisition of Technology from Multinational Corporations by Developing Countries" (New York: United Nations, 1974).

17. Ibid., p. 3.

18. A. S. Bhalla, ed., *Technology and Employment in Industry, A Case Study Approach* (Geneva: International Labour Office, 1975), pp. 6-7.

19. White, "Appropriate Factor Proportions," pp. 131, 135.
20. See Chapter 2 and Chapter 3.
21. See Chapter 4 and Eckaus, *Appropriate Technologies.*
22. See Chapter 2, p. 27.
23. White, "Appropriate Factor Proportions," p. 150.
24. See Chapter 2 for a discussion of this point.
25. White, "Appropriate Factor Proportions," p. 152; and Chudson and Wells, "Acquisition of Technology," pp. 7-10.
26. White, "Appropriate Factor Proportions," p. 153; and Bhalla, *Technology and Employment,* p. 7. Also see the reference to the bamboo tubewell in Chapter 2.
27. See Chapter 3 for a discussion of this point.
28. See Chapter 3.
29. Stewart, *Technology and Underdevelopment,* pp. 92, 103.
30. See Chapter 4.
31. Eckaus, *Appropriate Technologies,* p. 17; and Bhalla, *Technology and Employment,* p. 7.
32. See Chapter 2 on this point.
33. Chudson and Wells, "Acquisition of Technology," p. 20.
34. See Chapter 2 on this point.
35. Denis Goulet, *The Uncertain Promise* (IDOC/North America, 1977), pp. 81-82.
36. Ibid., p. 82.
37. Ibid., p. 69.
38. See Chapter 2 on this point.
39. See Appendix C for a list of participants at the workshop. None of the points in this consensus should be attributed to any of the individual participants at the workshop.

2
Appropriate Technology: Obstacles and Opportunities

Gustav Ranis

1. Introduction: The Problem and Its Context

Our initial task must unhappily be one of definition. So much has been said and written in recent years about "appropriate," "intermediate," "labor intensive," or "traditional" technologies—as opposed to "inappropriate," "capital intensive," and "modern"—that to present one's own interpretation of "appropriateness" before plunging into the subject becomes virtually obligatory. It would be comforting to say that the issue is wholly semantic; unfortunately, it is not. One of the reasons this topic is shrouded in so much controversy and confusion is this very lack of substantial agreement on what people are really talking about.

What we shall mean by an appropriate technology choice is the joint selection of processes and products "appropriate" to the maximization of societal objectives given the society's capabilities. This means that the appropriateness of technology choice includes product quality as well as technique choices and that they must be defined relative to the society's time-specific endowment as well as to its time-specific objectives. In other words, a society's endowment and its preferences, as among growth, the satisfaction of basic needs, and equity should give us a unique match-up with the "right" basket of devisable techniques and goods.

This seems a relatively tall order but it is really not too complicated, at least conceptually. An optimum choice can be established only if all the information is available and all the known choices are realizable. It is further simplified by the fact—only asserted here—that there actually need be no conflict between such objectives as growth and equity in the garden variety of labor-surplus developing countries. In other words, appropriate technology choice can make a large difference in improving performance in terms of employment, income, wage shares, and the family distribution of income.[1]

The real problem is not that our definition of "appropriateness" is wrong but that it is tautological. The proponents of "small is beautiful," of "secondhand is appropriate," of "labor-intensive is good," are able to make much stronger positive statements. Like the "big push" cultists of yesteryear, they know what the proper choices are. Their problem is that they may at times be wrong. We instead have opted for being tautological. What we are saying, to put the case more positively, is that the appropriate technology will vary across countries according to differences in endowment and tastes, as well as over time within a country. Moreover, the appropriate process for a poor labor-surplus economy is not always labor-intensive, and the appropriate good is not always a basic good. There is even less validity to the presumption that technologies appropriate to developing countries must be somehow "traditional" or, at best, "intermediate" in some sense of progressiveness or modernity.

There is an empirical presumption, supported by theory, that the more severe the population pressures on the land, the larger the population, and the greater the shortage of capital, the more likely it is that appropriate technologies will locate themselves at the labor-intensive and basic needs end of the spectrum. But all the empirical evidence we have come across would lead us to conclude that appropriate technologies are as likely to be "modern," machine-paced—and based on current vintage blueprints—as they are to be of the "traditional," "handicraft," or "secondhand machinery" variety. In other words, "advanced technologies" are not to be necessarily equated with the latest "advanced country" technology. They can be modern *and* labor-intensive, or modern *and* capital-intensive, use imported *or* domestic core technology, make use of extensive local adaptations *or not*. There are not easy com-

Appropriate Technology: Obstacles and Opportunities 25

fortable answers; it depends—on the place, the resources, the preferences, and the time. "Appropriateness" is little more than a useful tautology which sensitizes us to the existence of a wide array of technologies among which the one best suited to the particular circumstances can, at least potentially, be located or devised.

The problem is no longer, we believe, as it was a decade earlier, related to the failure to recognize the existence of such a wide diversity of potential choices; certainly not with respect to alternative processes. The literature now recognizes the existence of a wide range of alternative factor proportions for all but a small subset of continuous process industries. There have been a number of surveys[2] documenting the wide technological choices available in nature. These surveys agree that there are substantial substitution possibilities even with respect to the core process. In a wide range of industries, from sugar refining to beer to textiles to shoes, there may be as many as five major choices for even this core technology. Moreover, for example, by changing the speed of operations, the number of shifts, the intensity of the maintenance of the machinery—each of these core choices can be made substantially more flexible. Even greater opportunities occur in the peripheral production activities, such as transporting materials within the factory and storing and packaging the final product. The evidence indicates that supervisory labor to manage a large number of the unskilled along a machine-paced production line may be more readily available than the more highly skilled foremen required for highly automated lines.

However, such adaptations are not equally feasible in all industries. The closer the production process is to the raw material processing stage, the smaller the possibilities usually are for incorporating capital stretching adaptations. Conversely, the larger the proportion of processing done near the finished product stage, the greater the opportunities for labor-capital substitutions.

Finally, there is agreement that alternative organizational devices, including taking advantage of economies of scale at some stages of processing while subcontracting the other stages, can provide a substantial reduction in the costs of the plant, thus reducing overall capital intensity.[3] In addition to subcontracting different processes, the multi-product firm might also choose to subcontract product lines, especially to its subsidiaries. By decreasing the number of products or models produced in any one plant the firm is able to utilize existing production capacity more fully (because fewer machine changes are required), and to improve the coordination of plant production. Baily[4] found this to be true in the case of the Colombian shoe industry. Firms that specialized in fewer product lines were able to produce a greater quantity of shoes for a given plant capacity.

Efficient firm organization and management, while not always affecting relative capital-labor inputs, can result in greater output for given inputs. For example, if the firm is able to improve the flow of goods from one stage to the next or to coordinate storage facilities with deliveries, it will make better use of existing capacity. Santikarn's study of the Thai textile industry[5] found that its firms thought that Japanese firms had a competitive advantage, not because of superior technical know-how but because of better management—an important, sometimes hidden, "third factor."

The new conventional wisdom about the wide range of technology choices is less firm on the subject of the range of alternative goods within a narrow spectrum of Standard International Trade Classification (SITC) definitions. Following the theoretical work by Lancaster[6] and its application to developing countries, especially by Stewart,[7] our increased concern with the appropriateness of goods as a form of technology choice can be translated into the decomposition of a commodity into a bundle of quality characteristics. Planned—as opposed to accidental—variations in such characteristics can then be associated with substantial variations in the basic process choices open to the entrepreneur and to society.

It is curious that when an economist talks about technology choice he is usually referring to a process choice, i.e., choice of techniques, while the businessman is much more frequently concerned with small variations in product choice, i.e., choice of quality attributes. Our basic premise will be that at any level of SITC classification the assumption of homogeneity in the bundle of quality characteristics attaching to a particular good may obscure important additional sources of choice of which entrepreneurs can take advantage, in terms of both the utilization of the existing factor endowment and the provision of the so-called basic needs in the internal markets of developing countries.

We do not wish to be misunderstood in one crucial respect. It is not our view that there is some vast national and international shelf of technology ready to offer just the "right" process or product to be plunked down in a particular country and industry context. While the choice of a technology already in use elsewhere is a critical step—and by no means easy or costless, as we shall see—modifications will almost always have to be made before it can be installed and become fully "appropriate." Such modifications may constitute major technology changes in addition to, say, imported technology or minor adaptations in what has been or is being used or consumed elsewhere. The adoption of an appropriate process or good within a developing country always requires selection from the existing array of blueprints, plus major or minor modifications required to suit the unique local conditions. Each private or public entrepreneur makes both of these choices—often simultaneously—whenever a production decision is contemplated. In the process, he has at least potentially available a wealth of techniques for a given bundle of quality attributes as well as many attribute bundle alternatives.

Despite the wide potential choice in available processes and alternative commodity specifications, we are nevertheless confronted with the empirical fact that the selections actually made in most of the developing countries still appear to be substantially "inappropriate" by any known quantitative or judgmental standard. Capital-labor ratios differ between rich and poor countries for given standardly specified goods, but they vary much less than the well-known disparities in endowments would lead one to expect. Output mixes also differ, but not as much as resource gaps and the theory of international trade would predict. Certainly while the consumption per capita of the famous drip-dry shirts and Western shoes in the developing world is lower than that of bush shirts and open-toed sandals, the production and consumption of the latter items is less than might be expected.

These easily substantiated facts impress the observer. More interesting and more important, of course, is the explanation. A litany of causes, both domestic and international, is usually advanced. This litany often seems like the old-fashioned "factors in economic development" or "factors in the selection of appropriate technology" list. Any real effort to understand the problem of appropriate technology must at least order these elements in some meaningful fashion as a preliminary to the suggestion of internal or external policy actions. The obvious and insistent gap between textbook appropriateness and what one encounters in the real world, with serious consequences for both employment and income distribution objectives,[8] will be analyzed in this paper as a preliminary to the discussion of policy issues.

Our analysis of obstacles to the effective deployment of appropriate technology will be divided into two parts: Section 2 deals with the effective demand for appropriate technology as we have defined it within the developing countries and Section 3 covers the effective supply of such choices. The demand factors concentrate on the conditions at the individual entrepreneurial level, public or private, which provide the necessary incentives to search for a more appropriate set of processes and goods. This involves an examination of the market signals provided by the economic environment, which may or may not accurately reflect the basic resource conditions of the society. On the supply side we are dealing with the actual availability or absence of theoretically obtainable alternatives and the human and institutional capabilities that a society needs to discern and develop them. Obstacles to the effective working of such a technology market, in both its demand and supply dimensions, will be examined as a preliminary to the conclusions for policy in Section 4.

2. Weaknesses in the Effective Market for Appropriate Technology: The Demand Side

The effective demand for appropriate technology has at

least two main dimensions: one is the amount of inherent pressure on decision makers to seek out the best techniques available; the other is the extent to which their private search for "appropriateness" coincides with or deviates from social "appropriateness." While the first dimension has been relatively neglected, the second has received a lot of attention from economists, under the general heading of "the impact of relative price distortions on technology choice." It may nevertheless be true that the first is substantially more important in the real world, certainly in the case of nonagricultural activities. Let us briefly examine each.

The typical developing country, as most readers will appreciate, emerged out of the colonial production and resources flow structure after World War II—somewhat earlier in the case of Latin America—with post-independence efforts characterized by an increase in protection and intervention by national governments. Specifically, instead of continuing to let primary export receipts be used for reinvestment in traditional sectors and service activities facilitating these enclave activities, governments instituted an array of primary import substitution policies, mainly intended to shift both foreign and domestic resources into the new, rapidly growing consumer goods industries supplying the domestic market. When this subphase of development ran out of steam, the majority of countries turned to what is usually called secondary import substitution, i.e., the replacement of previously imported capital goods and the further processing of raw materials. Only a minority of countries, like Taiwan, took a different path at the end of primary import substitution. They continued to produce and now export the same relatively more labor-intensive consumer goods previously supplied only to the domestic markets.

But the basic point here is that the policies of import substitution, primary and even more so secondary, are by their very nature intended to provide protection to the domestic industrial producers in their fledgling period, and even beyond. This translates into the dispensation of all kinds of noncompetitive favors, not just for protection from foreign competition, but in terms of providing an oligopolistically or at least monopolistically competitive situation in the domestic industrial sector. Windfall profits are thus guaranteed through the import licensing system, with overvalued exchange rates, artificially low interest rates to favored borrowers, the direct allocation of strategic materials such as steel and cement, and the whole litany of favors well known by now. All these measures are aimed at ensuring safety and profits to the new class of industrial entrepreneurs, whether in the public or private sector.

There is no need for one more critique of the policy of import substitution and its effects, especially a policy that is maintained inflexibly and too long. But the point is that it creates a lack of competitive pressure, especially in nonagricultural pursuits. When an entrepreneur is guaranteed profits of 25 percent or 30 percent just by being at the head of the queue for whatever "goodies" are being allocated, his impetus to seek out the most appropriate technology is severely blunted. It is not part of the profit-maximizing religion of economists, but nevertheless observable, that entrepreneurs are much less anxious to increase their profits from 20 to 30 percent than from 10 to 15 percent, assuming that they have been guaranteed their basic cushion by virtue of the monopoly windfalls granted by government action. This satisficing behavior or preference for the "quiet life," as it is sometimes called, is also observed in advanced countries, at least in some industries.

The prevalence of satisficing behavior of this kind, when people's energies are much more profitably expended on seeking government favors and ensuring the maintenance of their monopoly position than on the search for more appropriate goods and processes, cannot be underestimated. The reason it has taken economists so long to give this dimension its proper weight is, I believe, the fact that it is relatively less important in the agricultural sector, which remains dominant in most developing economies. The usually small-scale, peasant, atomistic nature of the agricultural sector in developing societies makes this particular inducement to inappropriate technology choice less important. The typical farmer is normally "up against it." He is highly risk averse and depends on the government to convince him that new agricultural technology, more often produced in the public sector, is something in which he should be interested. It is not the exercise of his market power that leads him to reduce his effective demand for appropriate technology.

We turn now to the second dimension of the impact of the import substitution policy syndrome on the effective demand for appropriate technology. Most of this area has been pretty well dealt with in the literature, with one notable exception to which we will refer. The basic proposition here is that the thicker the veil between an economy's equilibrium or shadow prices for factors of production, as well as for finished goods, and the prices which appear in the market as a consequence of various kinds of government interventions, the more distorted the technologies that actually emerge from the selection process. The frequent overvaluation of the exchange rate cheapens capital goods and imports; minimum wage legislation, protection of workers against dismissal, the encouragement of unionization, and a long list of social legislation artificially raise the price of unskilled labor; low interest rate policies and accelerated depreciation allowances again bias entrepreneurs in favor of the use of capital; price controls on basic consumer goods, as well as such important producer goods as cement and steel, lead to further distortions in terms of quality as well as technology choice. The list is considerably longer. The distortions seem, moreover, to be more pronounced for larger firms both in terms of the relative impact of credit rationing and of the reach of minimum wage legislation, for example. There can also be little doubt that the distortions are more severe in some country contexts than others, depending on the inflationary pressures and the relative flexibility or rigidity of the protective measures, e.g., as between the use of tariffs and quantitative controls. Such differences among protective regimes are naturally important for the technology chosen.

One aspect of the import substitution policies on effective demand for appropriateness in technology has been given less attention than it deserves. That is the discouragement of the production of appropriate goods as a consequence of the relative discrimination against agricultural output during this policy regime. We have already noted the usual effect on agriculture's terms of trade; but we have not commented on the fact that the resulting lowering of agricultural incomes, coupled with concentration on the substitution of previously imported consumer goods by domestic imitations, has the serious consequence of stifling the demand for basic goods, i.e., goods with a more appropriate bundle of quality characteristics. During the colonial period in the Philippines, to cite but one example, imported consumer goods managed to kill off much of the efficient handicraft industry, and when import substitution took over in the post-war era, textile manufacturers in Luzon were in a position to compete effectively for what little demand there might be in the outlying areas of Mindanao. The big problem, of course, was the lack of market demand in the absence of balanced growth in the rural areas. This is quite in addition to the fact that whatever demand existed was being absorbed by inappropriate, or over-specified urban goods.

Traditional industry can often provide the base for modern industry. Building on local appropriate goods and technological traditions may seem rather primitive, but in many instances it really constitutes rebuilding traditions destroyed by a tendency to "import reproduction" rather than "import replacement" and by a government's frequent prior willingness to destroy village industries. An interesting example may be the plastic sandals factory, introduced to provide low-cost sandals with the effect of destroying both the local sandal makers and the local leather industry. Another may be the large-scale bread factory (in Dar-es-Salaam) which put local bakers out of work while extra flour had to be imported to supplement the local supply and, to top it off, the factory was not running at full capacity.[9] Once again, we are not arguing for the preservation of outdated handicraft industries but for the realization that such industries may well provide the core for modernized competitive production functions.

The strong underlying demand for more consumer goods in developing countries means that there is no lack of effective demand, in the advanced economy sense, for final goods. But a growth strategy is called for that is sufficiently participatory in character to satisfy basic demands across the entire landscape of the typical developing country. Appropriate goods and processes must be made available, not only to provide the satisfaction consumers are seeking but also to permit a greater utilization of unskilled labor, the typical developing economy's abundant resource. One consequence of the typical, narrowly-focused, urban-oriented, import-substitution growth path is the suppression of a potential domestic market for appropriate goods. Thus, still on the demand side, entrepreneurs are likely to be fixated on the demands of the enclave industrial urban elite as well as, in some cases, of export markets. Once again, the effective demand for appropriate technology, as we have defined it, is weak as a consequence of the growth path which has been chosen. Given a relatively stagnant agricultural sector, we cannot expect rural industries and services to grow, whether they constitute inputs into the agricultural sector or outputs demanded by agricultural families out of their slowly growing incomes. We have in mind agricultural implements in the first category and locally manufactured textiles in the second; one example is the spectacular growth of rural engineering industries linked to pump and tubewell production around Daska in the wake of Pakistan's Green Revolution–generated agricultural spurt of the early 1960s.

All imitations of Western goods are, of course, not necessarily inappropriate. Secondhand cars, tin roofs, and transistor radios may be reliable and relatively easy to maintain, and they may meet specific needs. The appropriate good may be cheaper in terms of its capacity to embody relatively low-cost primary inputs (e.g., unskilled labor) without sacrifice in basic performance (e.g., standard vs. electric razors). Moreover, the appropriate good may have greater intrinsic utility. A historical example can be found in the 19th century Japanese cotton textile industry. When the Japanese introduced a lower count yarn they were, in addition to lowering their costs and producing a more marketable commodity, producing a more appropriate type of cloth. The new product was not only cheaper but also warmer, sacrificing strength, a quality dimension not prized by the consumer.

Some time ago, Schmookler presented a thesis in the context

of the advanced economy to the effect that the pattern of final demand is the main determinant of invention, i.e., that the demand for appropriate technology is likely to be a derived demand.[10] He found that the number of capital goods inventions in a given field tended to vary over time with the sale of capital goods, and that the number of inventions, as measured by patents, tended to vary among industries in proportion to the sale of capital goods. When supply factors are entirely neglected, he concluded that inventive activity would be highest where prospective sales and profits were greatest.

Schmookler was talking about final consumer demand rather than the demand for technology, but they are related, especially in the area of the appropriateness of different kinds of goods. The stagnation of rural incomes, for example, serves to curb market demand and thus stifle the effective demand for this particular kind of appropriate technology.

In other words, while the supply of knowledge may set the upper limit on what is possible and may provide the intellectual stimulus, it is demand and the expected financial payoff that determine the ultimate direction and timing of inventive activity. For example, as long as horses were commonly used there were numerous inventions to improve the horseshoe. It was only when the use of horses declined that the number of related inventions also dropped. Similarly, the patterns governing the quantity and timing of track- and non-track-related railroad inventions coincided despite the differences in cost and technical complexity of the respective changes in technology.

In summary, the demand for appropriate technology tends to be limited by the level of workable competitiveness in the developing society and by the specific impact of policies that distort signals among factors and commodities and disturb whole sectors. These factors also may be viewed as threats to the appropriateness of the technology choices actually made. Removing obstacles is a necessary but not of itself sufficient condition for achieving appropriate choices. As Rosenberg points out, "economic forces and motives do not act within a vacuum but within changing limits and constraints of scientific and technical knowledge."[11] If the required knowledge with respect to alternative processes or quality bundles is not available or if the capacity to modify what is on the shelf does not exist, appropriate technology cannot be selected, nor can innovations toward greater appropriateness be made. To complete our story of the effectiveness of the market for appropriate technology in the developing world, we must now turn to the supply side.

3. Weaknesses in the Effective Market for Appropriate Technology: The Supply Side

If we were to picture the full storehouse of human knowledge potentially available to a developing society today, we would, of course, include every known combination of factors to produce a given good plus every known alteration of a quality bundle, giving us a potentially huge number of production points on the assumption that there are eager and able takers reacting to a realistic set of signals. But we realize immediately that such a shelf really does not exist. There is a simple lack of illumination with respect to a technology which may have been in use locally, say, twenty years ago or even is used today in another country. In other words there is a remarkable lack of information in the hands of the individual decision maker; moreover, high search costs are involved in illuminating a wide range of points on the production possibility contours to be found in both the rich and poor countries.

In addition, there is a problem over the appropriability of some of the information due to imperfect competition on the side of the technology seller. This applies to both processes and goods, but more strongly with respect to product or quality differentiation. This makes information practically unavailable or slows its diffusion. Under the cloak of patents, licensing, trademarks, etc., potentially available technology choice points are not always visible to all takers.

The record of foreign investors in this context is very much in dispute. They are supposed to offer a superior capacity to scan across countries; yet they are suspected of misusing patents, trademarks, and licensing arrangements by tying up technology with other items of transfer, rather than choosing the processes and goods most appropriate for a particular host country.

There has been much debate on the role of patents in technology transfer and indigenous technology change. One side holds that patent production is necessary to encourage domestic industrial R&D, and that the granting of patents to foreigners is necessary to induce the transfer of technology and capital. Opponents point out that patents granted by less developed countries (LDCs) to nationals generally represent only 10 to 25 percent of the total and that patents awarded to foreigners, mostly large transnational enterprises, often fail to yield the expected technology and investment transfers. There is substantial evidence that often patents registered by foreigners are not really utilized, but represent a device to protect markets and restrict entry for both local and competing foreign technology suppliers. Even when patents are "worked" in the LDC they often do not bring the expected new flow of investment. Vaitsos, for example, cites the case of Colombian pharmaceuticals in which the granting of patents not only failed to promote foreign investment, but was used to force the sale of local firms to transnationals.[12] All of this is additional to the problem that technology transfers themselves may not be conceptually or financially separable from other parts of the foreign investment bundle, for example, the movement of capital, management, and technical assistance.

Within the developing country, moreover, both the human and institutional capacity must be in place if one is to make appropriate shelf choices. Even if the demand conditions are approximately right there is a question of being able to discern, from the full storehouse of technological information, the proper combination to fit the particular local economy. We hear all too often of the neglect of "third factors" or overheads which make the transplantation of particular pieces of foreign technology highly inappropriate. At best, the attempt to re-create a foreign production function within a developing economy may lead to a substantial period of so-called x-inefficiency before lost ground can be recovered.[13] At worst, such cases of "turn-key projectitis" lead to the permanent inefficiency and inappropriateness of the process or product.

It should be emphasized again that this represents a human and institutional failing, i.e., on the supply side, as opposed to the imperfect competitive and market distorting features on the demand side. Admittedly, however, they both may be in evidence at the same time and prove difficult to disentangle in any real world situation.

The institutional capacity involved here entails the ability not only to make the right choice but also to diffuse a given imported or, for that matter, domestically selected technology. Diffusion channels require institutes, salesmen, the private banking system, government extension services, catalogues, fairs, etc.; and they are not a trivial matter. These networks do not necessarily exist in many developing countries, and when they are available they can be strongly biased or inefficient. This observer, for example, has noted the use of a highly efficient bamboo tubewell in one part of Bangladesh that is not known in very similarly situated regions in other parts of the same country.

Many of the so-called science and technology institutes one does find in the developing countries are often concerned with other matters. If government policies are designed to encourage firm-level R&D, the large-scale firms seeking advanced technology will tend to benefit. They would more likely have the financial resources needed for such activity, along with a greater risk-taking and absorptive capacity. Even government-financed research shows a bias in favor of large-scale advanced technology. University-centered engineering research, which probably would be government-funded, tends to emphasize the frontiers of scientific knowledge. Both the university reward system and the siren song of the "international college" of scientists and engineers naturally reinforce this bias. Professors are awarded grants and given promotions for enhancing the most advanced techniques, not for making existing techniques more appropriate. Research institutes also tend to favor advanced technology, again because of the prestige involved, and because of the general, if silent, conviction that this is where the most good can be accomplished. Similarly, most graduates are employed by large firms or, in the public sector, on large-scale projects working with relatively modern technology. While we by no means wish to imply that only breakthrough technology can be taught in the university setting, the existing incentives usually produce that outcome.

The ability to ask the right questions entails more than scanning international reference services or lists of technology alternatives. Knowing one's own situation specifically enough to request what is really needed is the essential element, and this is not a common resource, either from the individual or institutional point of view. It requires a minimum scientific and technological capacity, not necessarily based on high levels of educational attainment but on the kind of technical literacy many societies may not yet have. Asking the right question is the best protection against being given the wrong answer.

A very important dimension of this same inadequacy, which may be an important supply constraint, is the lack of recognition of the quantitative importance of existing trade-offs among attribute bundles for a given commodity. This could result in very different relative factor use outcomes within a given local demand context. International demonstration effects, both in terms of technologies and internationally advertised goods, may act as depressants on the effective demand for appropriate technology; they frequently contribute to imperfect competition associated with the pursuit of prestige technology and prestige consumption patterns; and they increase the difficulty of responding adequately to the demands of local public and private entrepreneurs. But identifying these performance characteristics seems less important than realizing the basic nature of the obstacle depicted here. That is, preconceived notions of technology or foreign-generated inappropriate tastes will bias the consumers of both technology and goods toward inappropriate choices.

Other supply constraints should be mentioned. We have already referred in passing to the monopoly power frequently exercised by transnational enterprises. This is especially visible in the case of product quality differentiation in the consumer goods area where technology is substantially appropriable, and where the technology being made available to particular developing countries may be subject to market sharing arrangements and be influenced by the response of the transnational enterprise to its own competitive pressures. Other restrictions appear in the form of tied aid in the public sector. These inhibit the free choice of technologies, mostly on the process side, in terms of the countries of origin for machinery and technical assistance. In other words, all the supply bottlenecks are by no means indigenous to the developing country; many have their origin in the rest of the world with which the typical developing country has to deal, whether or not they are part of the bundle involving the capital good import proper.

The Japanese seem to have done better on several counts here. For example, they used a lot of "reverse engineering," a process whereby they imported capital equipment and dismantled it to learn how it operates. Rather than reinvent the wheel, Japan thus learned to take advantage of its position as a late developer. The Japanese also encouraged the joint venture form of transnational enterprise (TNE) presence in their country from an early date.

Finally, and most important to this observer, providing an effective supply of knowledge to the developing society requires a recognition that no piece of technology on the shelf internationally, or for that matter regionally or even nationally, is ever quite "right" and ready to be put into place in some given context. While differences in resources and other environmental conditions may be minor, especially as we go from one region of the same country to another, I am very much impressed with the importance of human and natural "third factors" which can never be fully taken into account in any engineering production function. As a consequence, the ability to make minor adaptations on what already exists somewhere in the storehouse of human knowledge is of extreme importance. Such capacity may be found in the repair shops of the large company, in R&D institutes, or in the much less romantic confines of village blacksmith or bicycle repair shops. The significance of having a technical problem-solving capacity, often starting with the repair and maintenance functions and moving from there into adaptations and innovations, cannot be exaggerated. Without this capacity, modernized local versions of secondhand machinery or of foreign technology imports cannot hope to get off the ground. With it, the range of potential usable points in the production possibility set is substantially expanded.

We are aware that selection and adaptation often go hand in hand, but it is important to differentiate them conceptually. It is also necessary to recognize that the human institutional capacity to make choices from a static array of possible technologies and the ability to make requisite adaptations are not necessarily identical.

This adaptive appropriate technology capacity is related in very complicated ways to a country's scientific literacy. As pointed out by Kuznets and others, the causal chain does not always run from science, which is a universal good, to technology, which is more of a national or even subnational good; but it may run from a problem that needs to be solved at the technology level back to science. The experience of international agricultural research in recent years, via the so-called Green Revolution, is very instructive in this sense. We all know about the great scientific progress based on the genetic advances of Mendel, exemplified in the dwarf varieties of rice at the International Rice Research Institute at Los Baños. But it has also become increasingly clear in recent years that an adaptive national technological capacity is needed if the Green Revolution is not to run into serious difficulties; the moisture, soil, and temperature conditions are never quite the same in two given country situations.[14] The same point, if to a lesser degree, applies to industry, as demonstrated in the effects of humidity on cloth production, of pasture conditions on leatherware, and of the chemical quality of the water supply on plastics.

4. Conclusions for Policy: Strengthening the Market for Appropriate Technology

If the above general diagnosis is correct, certain policy conclusions follow. Perhaps the most telling is the extent to which the strengthening of both the demand and supply sides of this particular market depends on decisions within the developing

Appropriate Technology: Obstacles and Opportunities 41

world. The venerable notion that the technology leaders of this world can help the poor quickly, effectively, and (even better) cheaply, via the accelerated transfer of their superior technology, is quite off the mark—certainly in the simple and naive form so frequently encountered. Whether we are dealing with public sector transfers of technical assistance or private sector market transactions, appropriate technology selections unfortunately do not come cheaply. They can, of course, be assisted substantially from outside the developing world, but not without a prior understanding of what the basic obstacles to a better functioning of technology markets really are. In its absence there can be little doubt that, even with the best of intentions, help from the outside is as likely to contribute to the problem as to its solution.

What then is the most helpful approach to be taken from abroad? Not surprisingly, such an approach can only be effective when it recognizes what LDC policies and actions are required to remove—or at least reduce—the obstacles in the way of an effective demand for and an effective supply of appropriate technologies. Only in that context is it sensible to consider how the United States or any other outsider can really be helpful—if asked.

In the conclusions for policy which follow, we try to differentiate between domestic and international actions, and between actions directed at unleashing the forces of demand and those focused on eliciting an adequate supply response.

A. Suggested Demand Side Actions

1. *The primary responsibility for creating a competitive environment conducive to the creation of an effective demand for appropriate technology must lie with the LDC government.*

The demand for technology is most clearly a function of LDC governments' macroeconomic policies. The greater the flexibility of the virtually inevitable import substitution regime and the shorter its life, the less severe the costs of misplaced entrepreneurial attention and initiative. The greater the domestic freedom of entry into industry, for example, the less serious the fallout from the "quiet life" financed by windfall profits. And, of course, that holds as well for exclusive market arrangements and no-export clauses asked for, and bestowed upon, transnational enterprises. In brief, the more "workable" the competition in this phase of development, the lower the costs in terms of the foregone active search for appropriate technologies.

Specifically, the LDC government should seek to limit monopolistic and oligopolistic control of an industry. This can be done by patent reform to shorten the time a firm has exclusive control over a specific product or trademark, and by encouraging the production, by new domestic or foreign firms, of similarly specified products. To facilitate this process, direct entry restraints should be reexamined and special government credit arrangments should be established to help new firms enter hitherto closed-off industries. The main point is that governments can increase the extent of workable competition in a number of areas and thus place greater pressure on the successful search for more appropriate techniques and commodities.

2. *The main additional thrust of LDC macroeconomic policies should be to decrease the relative price distortions in various factor and product markets which currently obstruct the demand for appropriate technology.*

LDC regimes which have gradually liberalized and moved toward a more market- and export-oriented type of growth path have exhibited a much better record on appropriate technology choice than those that have chosen to retain an increasingly heated import substitution hothouse.[15] The drawing of a veil between endowments and factor prices, and between the availability of various types of intermediate and final goods and the administered prices established for them, inevitably has a negative effect on the selection of appropriate technologies.

To the extent the infant industry argument has validity—and we believe it has—the efficiency of the industrialization pattern and of the technology chosen may have to give way for a time to getting the job done and permitting the nation's new industrialists to learn by doing. While the distortions introduced by artificially low interest rates, overvalued exchange rates, etc., are frequently a necessary part of this early import substitution subphase of development, the critical issue really is the flexibility with which this policy regime is established and its duration. The critical question then becomes whether, at the end of this period, the political possibility exists for a gradual liberalization of the various relevant markets which would lead to a better static choice of technology and a better direction for technology change.

3. *The LDC government can generate additional demand for more appropriate goods and more appropriate technologies by aiding the balanced development of rural industry and agriculture.*

A balanced rural development plan, focusing on both agriculture and decentralized industry, is needed to ensure a steady domestic demand for appropriate technology. The agricultural sector, as is well known, has frequently received too little attention in past development efforts, especially during the primary import substitution sub-phase. It is important to avoid continued distortions in the internal terms of trade against agriculture, since we attach great importance to increased rural purchasing power as a precondition for balanced rural growth, i.e., the rapid expansion of both industrial and agricultural appropriate goods production for local rural markets. The more severe the distortion of the terms of trade against agriculture, the more serious the lack of purchasing power in the rural areas as a necessary precondition for such balanced growth, and the more unlikely the selection of appropriate techniques and commodities.

Optimal development requires the evolution of a strategy of the spatial expansion of industries, especially in activities not subject to pronounced economies of scale, keeping in mind the extent of dispersion of the population, for both market and labor supply reasons, as well as the transport system. The reallocation of government overhead investments toward the rural areas thus assists (rather than, as is so frequently the case, inhibiting), the spread of internal markets. It can be of the greatest assistance in the development and diffusion of appropriate technologies and appropriate goods.

4. *The LDC government should recognize that it is the trend of policy changes in the directions indicated that is im-*

portant—*not any unrealistically rapid shift in policy regime.*

Since risk aversion is a pronounced factor among most LDC private sector actors, a gradual but consistent policy moving in the desired direction is to be preferred over a wildly oscillating one that may be "better" on average. We recognize the substantial political as well as economic obstacles in the way of effecting transitions from more control- to less control-oriented policies. Such changes usually have to overcome strong vested interests in the industrial entrepreneurial community and the civil service. Therefore, the direction of policy is more important for entrepreneurial planning and the search for more appropriate technologies than the achievement of any textbook ideal. Oscillations of policy cause too much uncertainty and probably yield the worst results in the mixed economy setting.

In summary, LDC governments must take the lead in improving the demand conditions in technology markets. A strong, effective demand for appropriate technology requires both a shift from satisficing to maximizing behavior, and a reduction of the veil between market and shadow prices. Without these changes, there is very little chance that a decentralized industrial structure, utilizing appropriate technologies and appropriate goods, will make an appearance. A gradual movement toward a less protected environment and more realistic relative factor and output prices seems to be a necessary, if not sufficient, condition for the achievement of growth with equity in the typical LDC. In the absence of marked improvements in these conditions affecting entrepreneurial demand, "direct actions" on the supply side to induce a different growth strategy and a more appropriate use of technology simply will not work. The pace cannot be forced by pressure and laboratory demonstrations on the supply side.[16] In the typical, mixed developing economy, in any case, if the millions of individual decision makers don't "have the bit in their teeth," little else can follow. This "necessary but not sufficient" quality of the demand side of appropriate technology markets is well captured by Peter Timmer's bon mot that "getting relative prices right is not the end of development, but getting them wrong usually is."

Appropriate Technology: Obstacles and Opportunities 45

5. *DC governments can provide useful advice, technical assistance, and financing to help strengthen the demand side of technology markets, if they are asked.*

Strengthening the demand for appropriate technology is, as we have seen, largely a matter for domestic decision making and domestic action. Foreigners can help by providing advice and technical assistance, when asked, and by supplying capital resources to ease the country through the inevitable policy reform period.

By the same token, actions taken to provide substantial amounts of foreign aid are unlikely to be neutral with respect to the recipient country's choice of policy objectives. Such aid either helps to maintain or else can be used to change these objectives. Unless aid is minimal in size, therefore, it is more likely to be a question of the way in which foreign influence is brought to bear, i.e., the *style* of "consultation" or "intervention" rather than its absence or presence.

6. *DC governments can have an additional effect by their influence over private overseas investors who receive subsidized investment guarantees and other support.*

Before favorable action is taken by the Overseas Private Investment Corporation (OPIC), TNE sensitivity to the role of export prohibitions, the search for exclusive markets or market sharing arrangements, as well as other, additional protective legislation must be assured.

We cannot, of course, expect a U.S. or other transnational enterprise to be concerned with other than profit-maximizing criteria in any particular country situation. In that sense, at least, the large foreign subsidiary is no different from the large domestic firm, though it may have a shorter time perspective, given the higher real or imagined risks it is contending with, and an even greater preference for policy stability as opposed to its "average quality." But we can insist that public funds not be deployed indirectly to oppose the trend of policies agreed to by the LDC and donor governments.

In recent years, much of the attention of economists has focused on the above-mentioned market imperfections: on why price distortions have occurred; how they affect the economy; and what policies can reduce their negative effects. To a lesser extent the importance of strong competitive pressures has been emphasized. But so far much less attention has been given to the supply side of the technology market, as economists have traditionally tended to leave this dimension to technologists, scientists, and transnational enterprise executives. Unfortunately, in this market, supply does not create its own demand, nor does demand automatically elicit the corresponding supply.

We turn, finally, to the policy implications of our assessment of the obstacles to the effective supply of appropriate technologies.

B. Suggested Supply Side Actions

7. *Lack of information is one of the most serious obstacles to the effective supply of appropriate technology options. The LDC should develop a more effective network to disseminate existing relevant information.*

The gaps in the range of information actually available to the individual economic agents are enormous. This problem is not just due to the inadequacy of the book of international blueprints available to the typical LDC industrialist; more telling is the lack of information sharing among similarly situated LDCs and even among different regions—or villages—within the same country. The bamboo tubewell example mentioned earlier illustrates how often it becomes necessary to reinvent the wheel, or do without. While similar problems obtain in both agriculture and industry, government research institutes and extension services have more to offer on the agricultural side. No equivalent industrial R&D and extension service exists within most developing countries. The problem is not, we believe, tackled effectively by concentrating on international reference or other types of "question and answer" services; rather, the basic networks within the developing countries themselves must be strengthened.

8. *The same networks that carry information about alternative existing processes and goods must also have the R&D capacity to produce new information by helping individuals effect the inevitable adaptations.*

An institutional network focusing simultaneously on static

Appropriate Technology: Obstacles and Opportunities 47

information and the capacity for R&D should be encouraged. It is highly unlikely, given the best information network, that a particular process or quality bundle attached to a commodity will ever be precisely "right" for a particular rural industry or local market. In our view, the kind of network that will effectively carry information must also be capable of adapting shelf technology—foreign or domestic—to local specifications.

A likely place to start is the existing networks within the developing countries. Either their energies should be redirected toward the more "mundane" tasks—and away from "breakthrough" or frontier science and technology efforts—or they should be assisted in creating new capacity focused on rather non-traditional areas. The same networks that carry information about the so-called shelf of alternative processes and alternative appropriate goods must also have the R&D capacity to help individual entrepreneurs, public or private, to make the necessary adaptations and help with the diffusion across the countryside. Information, adaptation, and diffusion networks are conceptually of one cloth.

One of the often overlooked casualties of a society that depends on the importation of hardware and installs it without much modification is the resulting loss of participation and learning by experimentation. A decentralized indigenous innovation and technology adaptation network is important for building confidence in a society's own capacity to make appropriate technology choices.

9. *The network should probably contain a few centers of excellence in major industries and a substantial number of multipurpose institutions, some independent and some attached to development banks or to the branches of a commercial banking system.*

A combined information, adaptation, and dissemination network is essential because the human capacities to collect relevant information and assist entrepreneurs in the adaptation of existing technology and output modification choices cannot be separated. Each developing country probably needs a few centers of excellence in particular major industries, on the Madras Leather Institute model, where R&D on appropriate technology can go forward at the national level, aided by international scientific and technological linkages. Such centers should be linked with multipurpose, local adaptation and dissemination centers, spread throughout the country, which receive information and support from many other sources, mostly national or regional.

Some of the national centers of excellence, incidentally, might well become regional with time—not by asserting their rights, but by virtue of their performance. This is perhaps a better way to proceed than to try to imitate on the industrial front what has been successfully accomplished in agriculture by the various international crop research institutes. Advice, encouragement, and seed money can be very helpful in building a regional dimension into national centers of excellence that are part of the LDC information, adaptation, and dissemination network previously described.

10. *The proposed technology information, adaptation, and diffusion network could require substantial initial public sector financing, but most of the funding could be obtained by gradually redirecting existing science and technology allocations.*

The total financial implications of the aforementioned network in a developing country are not unreasonable, and very few additional resources should be required. Most developing societies already have a substantial science- and technology-oriented institutional network. Such institutions, however, typically have shown only a marginal interest in the area of appropriate goods or appropriate processes. The total LDC research and development effort is pitifully small, as is well known, and it is directed largely toward breakthrough technologies and frontier-science-related activities patterned on those of advanced countries. The redeployment of financial and human resources to the diffusion of information on alternative attribute combinations and appropriate technology change possibilities could be of the greatest value.

11. *More attention should be paid to the organizational structure, financing, and reward systems of the various institutes within the national network.*

One way to ensure that reconstituted institutions of this

kind become part of an effective network as we have described it is to force them increasingly to pay their own way by replacing public subsidies with private contracts. Moreover, an effort must be made to reward the scientific and technological staffs of such institutes in relation to these new objectives rather than exclusively by the applause from the "invisible international college" of fellow engineers and scientists. Assuming that the demand conditions exist, it would take very little to shift people and budgets into the kinds of networks described. Where appropriate, the whole system could be linked directly to rural development banks or even inserted into the supervised credit functions of a commercial rural branch banking system.

12. *The encouragement of trade associations should be examined as a way of speeding the dissemination of relevant information to small-scale users.*

Where economies of scale and information channels are important, as they often are, the encouragement of trade associations on the historical model of the Japanese Cotton Spinners' Association or the more recent Japanese trading companies model should be considered. These devices can help medium- and small-scale businesses to achieve some of the real economies of scale in raw materials purchasing and marketing and to gain access to government subsidies and favors. In the Japanese case, for example, activities included everything from financing to joint purchases, research, and training, as well as the publishing of technical journals and the funneling of information from central science and technology centers. As the technological gap between Japan and the West narrowed, trade associations lost some of their usefulness, but they are worthy of our attention for LDCs still engaged in the growth process. In the Colombian brick industry, to cite another example, visits by producers to other brick plants were found to be the most common source of diffusion of technical information.[17] Of course, this approach will work better where some degree of "workable competition" exists, since firms with a real competitive edge because of "invisible" technology will be more reluctant to share such information. When an LDC moves into a more advanced subphase of transition, as

Mexico and Brazil have done, the trading company model becomes more relevant, with its capacity to reach out to a large number of small enterprises. This is especially true in relation to participation in export markets.

13. *LDCs should specifically reexamine any institutional and organizational restraints on the diffusion of appropriate technology within their own economies.*

Through such policy steps as the revision of often very lax technology transfer, licensing, and patent agreements, LDCs can speed up the diffusion of new technology across firms. The shift from mule to ring technology in 19th century Japanese spinning, for example, was almost instantaneous, as was the agricultural sector's diffusion of the Veteran Farmer's improved seeds. LDC governments are often mesmerized by an international brand name and willing to sacrifice would-be local competition that might produce a cheaper and more appropriate alternative product.

14. *LDCs would benefit from the recognition that really meaningful R&D often takes place in unconventional locations and that its encouragement represents one of the best ways to enhance the country's own problem-solving capacities.*

There is much evidence to indicate that many of the adaptive technology changes that have occurred in the more successful developing countries have emanated from the repair shop or the factory floor rather than from the formal R&D establishment. What makes them effective is more a question of systems and software, including the sociology of the firm and its ability to reward suggestions coming "up the line," than it is of specific kinds of hardware decisions. The intensive maintenance of machines frequently encountered among the better performing developing countries—historical Japan and contemporary Taiwan, for example—is often associated with appropriate technology innovations in a labor-absorbing direction. Interviews in such countries' export processing zones have led this observer to conclude, for instance, that the subsidiary often takes pride in improving the particular process that is initially assigned to it merely because the environment offers relatively "cheap labor." Within the multinational corporation this process often means overcoming headquarters rules of thumb set for the company as a whole. That obstacle seems to be more formidable in the case of a company with only one foreign subsidiary than one which has a substantial number of them to provide more experience with appropriate technology variance. The sharing of such organizational devices among firms could certainly be helpful without divulging proprietary secrets.

15. *The development of an indigenous capital goods industry, especially on the mechanical engineering side, should generally be encouraged as an additional source of potential innovative R&D capacity.*

Appropriate R&D activity could presumably include both hardware prototypes (based on the scaling down of advanced economy technology), and such software activities as are required for in-plant, floor-level adaptive technology change. A particular target would be the creation of an indigenous capital goods capacity, especially at the village level. Some of that capacity already exists in most country situations in the form of repair shops, foundries, village blacksmiths, etc. Such individually non-spectacular but quantitatively impressive places usually offer indigenous innovative capacity that could be substantially enhanced with minimal outside inputs. Industrial co-ops, for example, have aided village blacksmiths, as in the case of Sri Lanka reported by Jéquier.[18] Elsewhere, the appropriate linkages can be forged, preferably with the commercial banking system or with rural development banks, in contrast to separate agricultural and rural industrial credit structures. Such encouragement of a simple capital goods industry within the developing country is likely to be helpful not only because the industry itself is normally labor-intensive at least in its mechanical branch, but also because it provides additional technological choices within each of the customer industries.[19] Nowshirvani has documented this process for the Iranian agricultural machinery industry sector.[20] There, many of the small-scale machinery producers started out as seasonal repair shops but, as their skills developed, they moved into spare parts and final products, making the considerable effort required to standardize and simplify capital goods production. Also, service guarantees were often necessary to attract customers.

Another effort to provide such standardization along with the production capacity is represented by the agricultural implements activity within the International Rice Research Institute (IRRI). Low-cost farm machinery was designed, based on the Japanese pattern of small, individually owned machines, and designs and advice were provided to any Philippine producer who was interested. The strategy was to encourage a decentralized, indigenous farm equipment industry. One firm was licensed to produce the power tiller transmission, since this was the most difficult task, while small, local metal workshops produced and assembled the rest of the components. The finished product was then marketed locally, usually at a lower price than comparable imports and providing considerable foreign exchange savings.

The production of machinery of a less simple type could in many cases be undertaken competitively, especially within some of the larger or more advanced LDCs. Capital-labor ratios in the machinery industry, especially of the mechanical type, are typically quite low, and the absence of substantial economies of scale makes the industry a natural choice. As Pack and Todaro[21] have pointed out, the experience of a number of countries has been quite favorable in this regard. An Economic Commission for Latin America study on Brazil showed that the prices of domestically produced machines and machinery components were below international prices, to cite but one example.[22]

16. *The DCs can contribute to the development of the LDCs own science and technology capacity by helping with the establishment of a well-functioning network within the LDCs.*

The developed countries can help to strengthen technological choice capacity within the developing countries by providing catalytic inputs, of the financial and human resources variety, for the internal network that has been outlined. Concentration on an enhanced capacity to ask the right questions and to choose more appropriately, rather than on transferring things, would be a wholesome change in emphasis in foreign assistance and other related efforts. Given our scientific and technological capacity, we can be especially helpful with the creation of the few national centers of excellence in the important industries relevant to a particular country or region. Policy advice and influence on the reallocation of indigenous budgetary and human resources, from the big breakthrough–big science emphasis to the above kinds of contributions along appropriate technology lines could make a substantial impact.

Special attention should also be given to encouraging domestic research in the area of appropriate technology. The existing institutional incentive systems reinforce the concentration on "frontier" research; but new priorities and an associated reward system could be devised to redirect some of the research. The DC can help here by establishing international awards for significant contributions to the development of more appropriate technology; encouraging the exchange of scientists; establishing a sabbatical leave system for LDC personnel; and supporting cooperative research activities among neighboring LDCs and between selected DC and LDC institutions.

17. *International reform of the patent system, trademarking practices, and other aspects of foreign investor behavior should be examined as a possible means of increasing the supply of technological information.*

The supply-side impact of imperfect competition in international markets should not be minimized, with respect to processes and especially with respect to products—as exemplified by patents, trademarks, licenses, and the like. We are not disputing the validity of some amount of protection, for a time, of the advanced country's own innovative talent; but the extent of socially necessary private appropriability of technological information is really much less than what is often encountered in practice in LDC markets. In this sense, international patent legislation, to cite but one example, could be usefully reviewed to ensure that patents are actually used (and that technology is transferred) rather than serving as a means of controlling markets and inhibiting the flow of information.

Current suggestions for the "unbundling" of various transnational enterprise packages to permit a more realistic and arm's-length transaction with respect to technology transfer could be very helpful.[23] The encouragement, via OPIC, of more flexible, time-phased contractual arrangements entered into by TNEs could make a contribution here.

It is entirely possible that transnational enterprises with unusual cross-sectional scanning capacities could be induced to exchange information on their appropriate technology adaptation experience across subsidiaries. This could be done in a manner that would not violate sensitive proprietary information constraints but, instead, emphasize principles, and thus give the transnationals a chance to act as catalysts and make a social contribution in an area where they often feel the subject of unwarranted abuse.

18. *DC bilateral and multilateral aid agencies should supplement their vocal support for appropriate technology by specific actions within their own lending programs.*

Public aid procedures and the influence of public or private capital movements often exercise a restrictive influence on the range of information and choice actually available to an LDC industrial entrepreneur. This does not mean that we believe in the feasibility of a perfectly informed, perfectly competitive, international aid and trading system. The fact is that the mystery and secretiveness surrounding the transfer of technology often is not warranted by the basic economics of the situation, but is instead the consequence of artificial and often misguided interventions by government as well as of unduly proprietary behavior on the part of private interests. It is well known that aid is often doubly tied, once to the country of origin, thereby restricting technology choice, and once again by the unwillingness of the donor to incur local costs, thereby eliminating the recipient country's own capital goods industry from supply consideration.

Aid agencies should be forced to seriously consider real world technology alternatives in their project lending, and a willingness on their part to permit local cost financing on projects would help reduce built-in preferences for capital imports. The adoption of such program lending, and flexible local cost financing rules, would increase the pressure for forcing consideration of technological alternatives within the cost-benefit analysis structure. This observer is skeptical about how many projects have in fact been approved on the basis of rate of return analyses rather than political considerations. But he is less skeptical about the potential for changing the technology of a project already agreed upon. The use, rather than discouragement, of local engineering consultants should be part of this effort.

On the private capital side, a similar willingness by government agencies to sensitize investors' technology behavior would be helpful, in return for granting subsidized OPIC insurance or contributing public funds to the investment itself. That would be among the actions that can be taken from the outside to assist in broadening the effective range of information and choice in an investment decision. The range of information and contact could well be broadened beyond an exclusive reliance on machinery salesmen fanning out from the former colonial mother or current donor country. Certainly this will occur if more third-country and internal LDC sourcing is encouraged in the inevitable discussion—now focused largely on financial matters—among DC embassy, OPIC, and TNE representatives. I am not suggesting a new DC control system; merely a widening of options by questioning the inevitability of the linkage between the source of the investment funds and of the capital equipment.

In sum, our analysis gives us grounds for considerable optimism as to what reasonably can be accomplished by policy actions, both internal and external, to render technology choices more appropriate. The concept is clearly multidimensional, and any search for simplistic or emotional solutions is likely to mislead us. Neither the return to Gandhian handicrafts nor the search for the big technology breakthrough is likely to prove particularly useful. The center of gravity of the appropriate technology concept rests, rather, with thousands of non-spectacular adaptive responses and modifications of modern processes and goods across a vast range of applications and landscapes.

The latest technology is not invariably inappropriate, nor

is the most basic invariably appropriate. A capital-intensive technology may be the most efficient, and poor people don't necessarily wish to buy expressly "poor people's goods." Cultural imperialism can alter taste preferences in techniques and commodities just as much as a policy of national self-reliance. And governments can always be expected to affect choices—e.g., against luxury goods—through tax policy. We certainly are not wise enough to preach on the exact nature of the socially optimal choices. All we would argue for is that LDC citizens be given the opportunity of choosing among alternatives, with the fullest possible information and with relative prices more adequately reflecting variations in the quality bundle. It is only in this way that more appropriate technology choices, which have both a societal endowment and an objectives dimension, have a better chance of being made.

Notes

An article based on similar material is expected to be published in Franklin Long and Mary Hughes, editors, *Values and Technology Choice*, Proceedings of an AAAS/Pugwash Conference, Johnson Foundation (Wisconsin, 1979).

1. The interested reader is referred to G. Ranis, "Development and the Distribution of Income: Some Counterevidence," *Challenge* (September/October 1977) and J. Fei, G. Ranis, S. Kuo, "Growth and the Family Distribution of Income by Factor Components," *Quarterly Journal of Economics* (February 1978).

2. D. Morawetz, "Employment Implications of Industrialization in Developing Countries," *Economic Journal* (September 1974); S. Acharya, "Fiscal/Financial Intervention, Factor Prices and Factor Proportions: A Review of Issues," IBRD, Staff Working Paper No. 183 (1974); A. Bhalla, *Technology and Employment in Industry* (ILO, 1975); G. Ranis, "Industrial Technology Choice and Employment: A Review of Developing Country Evidence," *Interciencia*, Vol. II, No. 1 (1977); Frances Stewart, "Technology and Employment in LDC's" in Edgar O. Edwards, ed., *Employment in Developing Nations* (New York: Columbia University Press, 1974).

3. The empirical reality behind these three loci of technology

choice is discussed extensively in the author's "Industrial Sector Labor Absorption," *Economic Development and Cultural Change* (April 1973).

4. M. A. Baily, "Technology Choice in the Brick and Men's Leather Shoe Industries in Colombia," AID (August 1977).

5. Mingarn Santikarn, *Technology Transfer: A Case Study*, Ph.D. thesis, Australian National University, 1977.

6. "A New Approach to Consumer Theory," *Journal of Political Economy* (April 1966).

7. Frances Stewart, *Technology and Underdevelopment* (Boulder, Colorado: Westview Press, 1977).

8. Issues which cannot be discussed here. See, however, D. Morawetz, "Employment Implications of Industrialisation in Developing Countries: A Survey," *Economic Journal*, Vol. 84 (1974); H. Chenery, M. Ahluwalia, C. Bell, J. Duloy and R. Jolly, *Redistribution with Growth* (Oxford University Press, 1974).

9. Both examples cited by Peter Meincke in "Appropriate Technology: Definition by Example," AAAS Conference, Johnson Foundation, Wisconsin, June 1978.

10. Jacob Schmookler, *Invention and Economic Growth* (Cambridge: Harvard University Press, 1966).

11. N. Rosenberg, "Science, Invention and Economic Growth," *Economic Journal* (March 1974).

12. Constantine Vaitsos, "The Revision of the International Patent System: Legal Considerations for a Third World Position," *World Development 1976*, Vol. IV, No. 2 (1976).

13. A concept based on the notion that firm efficiency cannot be determined by conventional production function analysis only, but that such "third factors" as employee experience, which evade quantification, may be relevant. Such factors in the transfer of turn-key projects have been labeled x-efficiency by some economists.

14. Robert Evenson and Yoav Kislev, *Agricultural Research and Productivity* (New Haven: Yale University Press, 1975).

15. For the contrasting experience of two developing countries, see the author's "Appropriate Technology in the Dual Economy: Reflections on Philippine and Taiwan Experience," *International Economic Association* (Macmillan, 1979).

16. Any more than (as we now understand) farmers will seek out new seeds if food prices are out of kilter, or families (as we still don't seem to fully believe) will demand cheap contraceptives if their condition motivates them to have large families.

17. M. A. Baily, "Technology Choice."

18. Nicolas Jéquier, ed., *Appropriate Technology—Problems and Promises* (OECD, 1976).

19. For more discussion on the importance of the capital goods industries, see Howard Pack and Michael Todaro, "Technical Transfer, Labour Absorption and Economic Development," *Oxford Economic Papers*, Vol. 21 (November 1969).

20. Vahid Nowshirvani, "Production and Use of Agricultural Machinery and Implements in Iran: Implications for Employment and Technological Change" (ILO, 1977).

21. Pack and Todaro, "Technical Transfer."

22. Cited in Carlos Diaz-Alejandro, *Essay on the Economic History of the Argentine Republic* (New Haven: Yale University Press, 1971).

23. For more on the potentially helpful role of the transnational corporation see the author's "The Multinational Corporation as an Instrument of Development," *The Multinational Corporation and Social Change*, Louis Goodman and David Apter, eds. (New York: Praeger Publishers, 1976). A number of the observations in Chapter 5 (this volume) and its case studies are also relevant here.

3
Technology and Employment: Constraints on Optimal Performance

Howard Pack

This essay will review two areas of major recent research interest and attempt to show their largely unexplored connection and its implication for the United States position at the United Nations science and technology conference. We first review some of the studies that have been done on the possibilities for using more labor and less capital in industrial production activities. As will be seen, the available evidence suggests a considerable scope for such substitution. However, for any individual firm or government enterprise, identification of appropriate equipment, linking a large number of independent pieces of machinery in a production unit, and finally, going through a shakedown period, may exceed its technical abilities. Put another way, even if factor prices facing firms were socially appropriate and firms wanted to use cost-minimizing production techniques, they might not have adequate managerial and engineering ability to realize this goal. We would then expect that if such skills are in fact important constraints to the adoption of appropriate technology, firms possessing these abilities would be more likely to choose it than firms in the same industry bereft of such skills.

It is, of course, not easy to define or measure the relevant cognitive abilities nor how their relative endowment varies among firms. However, it is widely asserted, and certainly seems plausible, that multinational corporations (MNCs) operating in less developed countries (LDCs) may employ or have inexpensive access to precisely the requisite skills while local competitors do not possess them. If so, the behavior of the two sets of firms should differ from that predicted solely on the basis of relative factor prices. In particular, MNCs should, *ceteris paribus*, exhibit greater labor intensity in producing a given product.

Part 1 briefly describes the current state of knowledge of the potential for substituting labor for capital in modern industrial processes. Part 2 considers some of the difficulties which may be encountered in realizing the feasible substitution even when it is privately profitable to do so. Part 3 describes and summarizes a number of studies of the factor intensity behavior of multinationals in LDCs. Part 4 sets these studies within a consistent theoretical framework and derives some policy implications.

1. Knowledge About Existing Substitution Possibilities

Though it is not our purpose to survey the literature on choice of techniques, a summary of some of the conclusions we have reached on the basis of recent literature is germane at this point.[1]

> A. In a large number of industrial activities a product is producible with a considerable range of alternate ratios of capital to labor. Much of the potential substitution of labor for capital stems from use of labor-intensive methods in "peripheral" production activities; labor, with little if any capital, can be used to transport material within the factory, to pack cartons, and to store the final product. The evidence for these statements is drawn from observation of both DC and LDC factory operations and engineering specifications.
>
> B. Evidence is also accumulating that the core production process itself, whether the cooking of food or the production of woven cotton cloth, offers efficient possibilities for using less expensive equipment and more labor per unit of output of *a given quality*, i.e., the more labor-intensive processes are not relegated

to the production of inferior commodities. Adaptation of existing equipment, for example, changing the "normal" speed of operation, offers still further opportunities to save capital and increase the relative use of labor. Finally, extensive underutilization of industrial capacity provides considerable scope for increasing the effective labor-capital ratio.

C. There are some industries that probably offer limited possibilities for altering the capital-labor ratio compared with that prevailing in advanced economies. These are typically activities in which most LDCs have no comparative advantage and where the basic problem is to forestall their introduction (typically behind tariff walls) rather than to suggest methods of changing production in a more labor-intensive direction. Even these classical no-technical-substitution cases offer some choice, principally in the price of equipment which typically varies among the sources of supply.

To establish some orders of magnitude a set of numerical results derived from a number of recent studies are shown in Table 1. These figures measure the ratio of fixed capital per worker necessary to begin a new production operation using varying *existing* technologies. The underlying engineering analyses have been carried out by teams of engineers and economists.[2] Variations in the capital-labor ratio reflect several factors: (1) different engineering principles used in the core process; (2) varying sources of supply of the same equipment as some firms or countries exhibit lower prices fairly consistently; (3) differences in the degree of mechanization of peripheral activities such as the bagging of fertilizer; and (4) minor adaptations in existing plants reflecting learning accumulated during the production process.

Thus, some of the most recent work suggests a rather substantial range of technical choice in modern manufacturing.[3] It should be noted that all of these studies consider substitution among plants of efficient size, in which no further scale economies are to be realized. The range of choice presented is

TABLE 1
Fixed Investment per Job Created (Dollars)

	Capacity per year	Alternative Technologies				
		1	2	3	4	5
Sugar refining	50,000 tons	2,869	3,480	3,869	5,987	6,204
Maize milling	36,000 tons	2,917	5,465	7,081	9,730	
Urea	528,000 tons	123,789	126,302	134,800	137,631	
Cotton yarn	1 ton	2,942	16,434			
Woven cotton cloth	1 million sq. yards	1,508	8,125	11,870	33,063	
Beer brewing	200,000 hectolitres	11,419	15,000			
Men's leather shoes	300,000 pair	760	1,042	2,157		

Source: Cotton yarn and woven cloth, H. Pack, "The Optimality of Used Equipment: Some Computations for the Cotton Textile Industry," *Economic Development and Cultural Change* (January 1978); all other studies at the University of Strathclyde; summaries appear in *World Development* (October 1977).

almost surely to be more circumscribed than the options to be derived from a still more systematic search that would include machinery produced in the LDCs themselves in the set of available techniques. For example, the engineering data for the textile study do not take account of the additional labor intensity afforded by semiautomatic looms produced in Korea. These are no longer produced in the industrialized countries but are currently in wide use in Korean textile production, including that destined for export.[4] Their inclusion would significantly widen the efficient range of choice in weaving.

If an LDC is planning the expansion of its industrial sector, its choices with respect to capital intensity are certainly not narrow. There need be no trade-off between output (in this case value added originating in manufacturing) and employment; as long as plentiful unskilled labor can be substituted for scarce capital both employment and output increase. The political authority has two dimensions of choice in view of the substitution possibilities. By providing a reasonably competitive atmosphere in product markets, particularly a liberal international trade regime, it will encourage the expansion of sectors which are not capital-intensive. The critical importance of correct sector choice can readily be seen in Table 1— compare the capital-labor ratio of the least capital-intensive fertilizer plant with that of the most capital-intensive shoe or cloth producing technology.[5] If policies are pursued, whether by direct government investment or by allowing distorted factor and product prices to prevail, which systematically encourage the expansion of sectors whose lowest achievable capital cost per job created is $124,000, only a limited contribution to employment growth can be expected from that sector. Though careful attention to design can reduce the cost per job from $138,000 to $124,000, the latter remains considerably in excess of the capital available to equip new labor force entrants.

The second choice facing the government, assuming that reasonable balance in sectoral expansion is achieved, is, of course, appropriate choice of technique within a given sector. Interestingly enough, the amount of direct evidence demonstrating inappropriate choice of manufacturing technology is remarkably small. Though anecdotes abound about automated factories employing very few workers, little is known about the relative importance of very modern equipment in a typical LDC's annual investment. The conclusion that inappropriate technology is widely utilized is based primarily on the discrepancy between measured growth rates of real manufacturing output and employment. This discrepancy may, however, reflect increasing capital per worker or greater labor productivity with the same capital per worker, or some combination.[6] We assume that at least some improvement in the employment-generating effects of new investment is possible; in the main, this assumption is based, not on inferences from the growth of average labor productivity in many LDCs, but on the level of exports of new capital goods from DCs to LDCs and the absence of any significant intra-LDC trade in capital goods, and the lack of substantial production except in a handful of countries.[7]

In addition, some of the microeconomic work on choice of technique, particularly that comparing local and foreign companies in the same productive activity, has indicated some inappropriate production methods. At least two studies explicitly addressing the question of whether firms, given their relative factor prices, have in fact chosen the cost minimizing techniques or a more capital-intensive one, have suggested that the latter is the case.[8]

2. Obstacles to Correct Choice of Factor Proportions

Assuming that private *and* public firms often choose an inappropriate technology where alternatives are available, can such behavior be explained? The most obvious, and until recently the most widely emphasized, explanation was that of incorrect factor market prices facing producers: the cost of unskilled labor being above and that of capital below their respective opportunity costs. While these are undoubtedly important elements, they are unlikely to constitute the sole explanation.

Consider a firm undertaking an entirely new venture or expanding an existing one. The profit augmenting returns to a search for appropriate plant and equipment must be weighed against the increase in profits to be obtained by alternative uses of entrepreneurial and technician time. Once the scarcity of both management and staff time (in the larger firms) is taken into account, it is obvious that many activities exist to which an allocation of time may be profitable. Better inventory control; improved floor supervision; searching for lower cost suppliers; identification of new markets; coaxing better treatment from government in the form of higher protection, rebates, or tariffs on inputs; and exploiting the entire panoply of government measures may be better uses of one's time than choosing "the" correct technique, though it too could improve profits.

These possibilities suggest that, all other factors being equal (the country, sector, product mix, and relative factor prices), the firms facing lower costs of acquiring the information needed for implementing a cost-minimizing, labor-intensive technology are more likely to take that step than companies for whom the (explicit and implicit) information costs are very high.

The adoption of a readily available capital-intensive technology may well be the profit-maximizing strategy for the latter. It is plausible to conjecture that MNCs have lower search costs than purely local companies involved in the same activity. They can easily identify and transfer equipment among subsidiaries, especially items that have become too expensive to utilize in higher wage countries because of their labor intensity. Indeed, the parent company may have established a new plant partly to utilize such equipment in the production of exports. Alternately, the local manager may ask the parent company's purchasing office for advice on the availability of used equipment if it is not available within the country. Given the well known difficulties of identifying reliable equipment,[9] such "costless" aid to the local manager (in both time and explicit costs) clearly increases the likelihood of his employing appropriate equipment in order to take advantage of the factor prices facing him. Not only can a purchasing office identify and physically evaluate the condition of the equipment, it also probably can obtain a better price, insofar as most studies of the used equipment market indicate that price setting in this market is one of bilateral monopoly rather than perfect competition.

Thus, if we observe the production of similar products in a country by both local firms and MNCs the preceding argument suggests that MNCs would utilize a more appropriate technology, given the relative factor prices facing the two types of firms. Put slightly differently, given the substitution opportunities in production, the MNC is more likely to approximate its cost-minimizing factor proportions than would a purely local firm.

This view, of course, contrasts strongly with the often polemical work that asserts, usually without evidence, that multinationals use excessively capital-intensive methods.[10] Insofar as the logic of the argument can be separated from the hyperbole of the typical discussion, it appears that the major assumption is that MNCs utilize the same production process in the host country as in the country of origin, since this is the only technology with which they are familiar, i.e., the search costs are less. However, this begs the question of why these otherwise profit-maximizing entities (clever enough to set transfer prices to maximize global net profits) engage in no profitability calculus on the benefits and costs of adaptation despite the widely differing factor prices prevailing in the home country and the LDC.

It is often overlooked that an MNC which has been in operation for some length of time is likely to have various vintages of equipment in its worldwide plants. As wages rise in the richer of the countries where it operates, the quasi-rents to be earned on older equipment will decline. Rather than sell such equipment in the used equipment market in the DCs, firms can earn higher returns by moving it to a lower wage location. It is considerably easier for an MNC than for a local LDC company to identify this equipment and purchase it directly from the vendor in the DC. Intracompany transactions in the face of great uncertainty about the quality of the equipment are likely to result in more movement of equipment than in a market in which one important attribute of the commodity, its quality, is very difficult to determine.

Thus, the possession of a considerable range of vintages within its international operations may permit adaptive MNC behavior with respect to technology. Clearly this cannot be the case when the underlying production process offers virtually no substitution opportunities, for example, oil refining and other continuous process industries. Though the multinationals do utilize older vintage equipment, there is no evidence to suggest that they engage in research specifically designed to modify production techniques to better suit LDC factor availabilities.[11] However, because most MNC production is in traditional branches such as food processing and hand soap, the use of traditional methods, twenty years out of date in the advanced countries, is almost surely the cost-minimizing strategy rather than the development of newer, labor-intensive techniques exhibiting improved productivity.[12]

3. Comparative Adaptive Behavior

We now explicitly consider some of the studies that have been completed on the factor intensity decision of multinationals. The analytic methods employed have not been uniform. Often they were determined by the process by which the data were generated. Thus, statistical tests have not been used in many of the studies, particularly those that describe actual production methods in some detail rather than simply relying on measurements of capital per worker. As will be seen, the available results do not support the hypothesis that MNCs are more capital-intensive than purely local companies despite the fact that they generally face a higher ratio of wages to capital costs.[13] Stated differently, they are more adaptive.

The studies that have been carried out can usefully be separated into two types: (1) those that rely on official published statistics or data gathered from firms solely from questionnaires; (2) studies that employ direct observation of factory methods to identify the technology in use as a supplement to the data used in the first set of studies. While one cannot just walk around in a factory and come up with an index of the technology utilized, it is possible to ask detailed questions about the source of equipment, when it was introduced in advanced countries, whether it is still in use there, and how current processes in advanced countries differ.

A. Statistical Analyses

Forsyth and Solomon consider whether capital intensity (measured by historical cost of fixed assets per production worker) is systematically associated with foreign, local, or resident expatriate ownership of firms in Ghanaian manufacturing.[14] Their principal finding is

> that nationality of ownership is significantly related to choice of technology, such that multinational corporations do indeed install plant and machinery embodying different factor proportions from those observed in domestic (LDC) factories. However, the direction of these differences is not always the same, and it is not the case, as has been suggested elsewhere, that multinationals always tend to be more capital intensive (or more labour intensive) than local competitors: this appears to vary from industry to industry.[15]

The authors arrive at these results using pooled industry

data. Their statistical method utilizes discriminant analysis which attempts to predict from which of the three ownership categories an observation is drawn, given the value of the variables characterizing it.

It is difficult to interpret their results within our framework. First, it will be remembered that we suggest that for a given wage-rental ratio the MNC is likely to exhibit a lower capital-labor ratio than does a purely domestic firm. However, Forsyth and Solomon in this paper do not explicitly consider the relation between the wage-rental ratio and the level of capital intensity, nor do the results they present allow inferences about it. Rather, both the wage-rental ratio and the capital-labor ratio are used to predict whether a given firm belongs to a particular ownership group. Moreover, as they pool observations across sectors, intrinsic intersectoral differences in potential capital intensity account for an unknown fraction of their results. Thus, despite the fact that their conclusions do not support the hypothesis of a systematic association between high capital intensity and foreign status, not much more can be inferred.

In another study, again using data on Ghanaian manufacturing, Forsyth and Solomon analyze the factor proportions characterizing domestic and foreign-owned firms producing well-defined industrial products.[16] Hence, intersectoral differences in capital intensity are not masked as they were in their earlier study. The data consist of official statistics, carefully screened to weed out unreliable returns. The authors do not find consistent, statistically significant correlations between form of ownership and capital intensity; the relations differ by product and few are significant.

Thus, the two studies of Ghana by Forsyth and Solomon, though of considerable interest, do not allow any clear inferences about factor proportions and form of ownership. One difficulty with their work, as with the next several studies, is the failure to explore technology at the plant level. Without such information, most statistical measures seem to mask the fairly distinct patterns that emerge at a more disaggregated level.

Responses to a questionnaire were employed by Reuber,

Crookell, Emerson, and Gallais-Hammond in analyzing the adaptive behavior of 77 multinationals in a large number of LDCs.[17] Of 78 firms, 57 replied that they introduced production methods without any adaptation of the home plant's technology. It is difficult to interpret these responses for a number of reasons. The major difficulty is that such questionnaires are typically answered by front-office personnel with limited familiarity with detailed production issues. A soft-drink bottler who uses the same bottling machine as does the home-country factory but different washing and loading methods is likely to respond that the same method is employed. White's[18] description of a study by Boon[19] is salient:

> (Boon) describes an interview at an engine plant owned by an MNC in Mexico. At the beginning of the interview the management assures Boon that the Mexican plant uses exactly the same technology as that used in the parent plant in the developed country. But as the interview proceeds and Boon tours the factory, it becomes clear that in many respects the factor proportions are different. The main machinery processes are automated, but secondhand equipment is used. And all of the auxiliary processes, like packaging, handling, transporting, and storing, are done much more labor intensively.

The second difficulty with the Reuber et al. study is that no information is provided on the industry breakdown of different responses. While a failure to alter the technology in chemical processing would hardly be surprising, a similar failure in weaving would be less understandable. Hence, the Reuber study, despite considerable merits in other areas, provides little information on the issue at hand.

A number of other studies have also reported statistical tests of differences in factor proportions between local firms and MNCs.[20] However, the data chosen to represent production techniques almost surely obscure the relations in which we are primarily interested. Thus, one study uses horsepower per worker as a measure of capital intensity though it has long been known that this measure has little relation to either the value of capital per worker or to the complexity of the capital stock in use. Another analysis uses replacement value of capital, though a major issue is whether MNCs make better use of used equipment whose acquisition cost is lower. Still other questions important for a discriminating analysis are necessarily omitted, for example, the comparability of products. Though all of these analyses are suggestive, they do not move toward study designs allowing discrimination among competing hypotheses.

One of the more recent of these statistical studies by Lecraw does move in this direction, however.[21] Though also relying on reported measures of capital intensity obtained through a survey, Lecraw's analysis moves much closer to a study design allowing a resolution of some of the underlying issues. He obtained data from 200 manufacturing companies in Thailand including purely local companies, western MNCs, and MNCs based in other LDCs such as India and Hong Kong. Branch by branch it is found that, given the difference in factor prices faced by the three sets of firms, the two MNC groups exhibit lower capital-labor ratios despite facing higher wage-rental ratios. Of considerable interest is the finding that LDC-based multinationals are still more efficient than those from the advanced countries, exhibiting both the lower capital-labor ratios and social costs of production. Finally, while there are some problems with the level of disaggregation (for example, battery firms include both high- and low-quality production), it is still true that, in the smaller set of cases in which identical products are manufactured, the earlier results continue to hold.[22]

B. Process-Level Analyses

The study by Morley and Smith[23] is the first we shall review that moves downward from "front-office" responses or officially collected data to observations at the plant level of current production methods. The authors analyzed the adaptation undertaken by MNCs in their Brazilian operations relative to their home country techniques. The industries analyzed were mainly in mechanical engineering. The major finding is "a substantial modification of production processes by U.S. multinationals in Brazil. They use one-third to one-fourth as much capital per man . . . (and) . . . use fewer automatic and specialized machines than comparable establishments in the United States. They also tend to substitute labor for capital in what we have called the materials handling or support services of the production process such as inspections, production scheduling, inventory control and the like" (p. 30).

Though the study does not deal directly with the comparative behavior of domestic firms and MNCs when the two are producing the same product, it does provide another strand of evidence that the MNCs engage in technical adaptation. It is worth noting that the major reason given by firms for such behavior was that the Brazilian market is smaller than that of the United States. At the same time lower labor costs are not, in the interviews, accorded much weight. Though similar evidence is suggested by other studies as well, the scale issue is not well understood. The Morley-Smith results are partly a reflection of the fact that a large percentage of their sample was taken from those subsectors of the mechanical engineering sector exhibiting very large average firm size in the United States (tractors, diesel engines) and typically thought of as exhibiting substantial scale economies as compared with say, industrial machinery.

In a study of 42 manufacturing plants in Kenya, the author found extensive adaptation of production processes designed to take advantage of the lower cost of labor as well as adjust to smaller markets.[24] These adjustments occurred in a broad spectrum of industries, from paint production to the canning of fruit. Several types of adaptation were observed: (1) the central or core manufacturing process employed used equipment or machinery designed for lower volume; (2) peripheral activities such as material movement, packaging, and storing were performed manually. Computations indicate that the economic efficiency of the latter depends critically on the low scale of operations rather than on a low level of wages.

The adaptation to local conditions was considerably greater among subsidiaries of MNCs than among domestically-owned firms. This was attributable to a number of characteristics of the respective managerial groups. Local owners and managers entered production after earlier experience in retailing or wholesaling, whereas MNC managers typically had considerable production experience. While the former group did not understand how to "uncouple" the machinery packages suggested by capital goods salesmen, those who had production experience, typically in United Kingdom plants in the early 1950s, were adroit at identifying both individual machines that allow low cost production and areas in which machines were entirely redundant (such as packaging). Moreover, the MNC managers were aided by their knowledge of the new and used equipment market possessed by their parent company's purchasing office.

Two of the most suggestive papers recently published are those of Wells who analyzes the behavior of LDC-based multinationals.[25] They, as well as their DC counterparts, might be expected to possess the organizational abilities that would allow the identification, purchase, and installation of appropriate equipment. Wells does find that Hong Kong–based Chinese firms are quite adept at identifying and implementing appropriate production methods when establishing joint ventures in Southeast Asia and elsewhere. Partly, this adaptive behavior reflects the transfer of equipment, initially used in Hong Kong, that is no longer profitable to operate given the higher level of wages currently prevailing there. The technical adaptation also is attributable to the production experience of, for example, Hong Kong's textile producers, whose familiarity with a variety of operating techniques reflects their origin in Shanghai in the early part of the century. By itself, access to appropriate machinery would be insufficient without the long history of efficiently utilizing a variety of such vintages in actual production. On the other hand, given the imperfections of the used equipment market, especially the difficulty of evaluating quality, the information network within the firm that allows it to evaluate equipment is also important. These observations parallel those made in Kenya on the importance of both previous production experience with older vintages and of the ability, in Kenya, of western-based MNCs to obtain appropriate equipment from other parts of the corporation. Finally, Wells notes that much of the adaptive performance consists of adjustments to the smaller scale typical of LDC markets as well as the occasional design of special equipment.

The question of the ability of LDC multinationals to compete with the western-based ones is also addressed. The latter, though they exhibit more appropriate factor intensity than purely local companies, are found nevertheless to establish plants that are too large for the small local markets. These plants run at a fraction of full capacity and exhibit higher costs than the more appropriately scaled LDC multinational plants.[26] The relevance of scale adaptation is supported by preliminary evidence of Balakrishnan, cited by Wells, that "Indian foreign investors tend to be firms with a plant size at home similar to that required in the foreign market. Other firms in the same industry, but with much larger or smaller plants in India, do not establish plants outside India."

4. Interpretation and Policy Issues

A. Interpretation

Although the preceding survey of the success of foreign-owned firms in adapting technology to local conditions may at first glance exhibit dismaying variety, I nevertheless think that the evidence can be interpreted to yield a fairly systematic picture which confirms our hypothesis of the importance of organizational abilities in addition to correct factor prices, if appropriate factor proportions are to be achieved. First, it is necessary to reemphasize that some of the studies, though interesting in other dimensions, do not explore the questions with which we are concerned. In particular, those such as Reuber et al., which simply ask MNCs whether they have adapted their production to LDC conditions but do not inspect the factory itself, are not sufficiently discriminating for our purposes. As indicated above, the respondents to the questionnaire tend to be front-office personnel often unaware of the details of production. Moreover, these broad-based studies do not report results by industry and therefore one has no measure of the potential adaptation which might have been undertaken. Finally, there is no information on comparable local firms or the relative factor prices faced by the two sets of enterprises. We are not criticizing the failure of some of the authors to write another book or article than the one which was produced. Our remarks are intended rather to show why their results are not germane to our interests and cannot be used as counterevidence.

In the studies that do look at the question of local versus foreign performance, several strains dominate the discussion. First, examination of MNCs industry by industry suggests that they do not replicate their home factory's production technology; in particular, the need to adjust to smaller levels of output constitutes an important incentive to adapt production processes. Second, where explicit comparisons of the equipment used in production have been carried out, foreign firms (whether DC- or LDC-based) generally exhibit more appropriate techniques. In the purely statistical studies, except for Lecraw's, the findings are more varied, reflecting, one suspects, poor measurement of the relevant variables.

The two major results—namely, that MNCs do alter factor proportions and that they often appear to be better at this than local firms—imply that, although correct factor prices may be necessary for an appropriate choice of technique, they are not sufficient. Companies without the requisite knowledge base—that is, knowledge encompassing both the ability to identify relevant equipment and to successfully use it in production—may not be able to achieve as low a capital-labor ratio as factor prices would dictate. MNCs in Kenya, Thailand, and other parts of Southeast Asia seem to embody such abilities within the organization.

It is appropriate here to enter one set of qualifications. The organizational ability we have ascribed to the MNC but not to the local companies is clearly too simple: purely local firms in some of the middle income LDCs in Latin America as well as Korea and Taiwan have demonstrated some of these endowments, particularly the ability to identify, purchase, and operate labor-intensive plants. Indeed, evidence is accumulating that locally-owned Latin American firms and MNC subsidiaries (which have achieved a large degree of independent status) have not only purchased such equipment but have engaged in design activities to augment plant-level productivity.[27] On the other hand there is little evidence of such skill in most of Africa and some of Asia. If these tentative findings are confirmed by further studies, they suggest that, proportionately, the benefits to be derived from the transfer of the relevant skills, by whatever means, are likely to be greatest for the poorest countries. This may be one reason why the adaptive behavior in countries such as Kenya and Thailand, that are short of local engineering abilities, is greater than that in Brazil, for example, which offers a much greater domestic skill base.

Despite the considerable evidence of successful adaptation of factor proportions, one feature continually appears in the various studies, namely, that such behavior may be economic only at low levels of output. More technically, the implication of some of the studies is that the underlying production function is non-homothetic—higher levels of output will be produced more cheaply with increasingly high capital-labor ratios even at unchanged relative factor prices. Labor, at any rate unskilled labor, while enjoying a productivity advantage at low levels of output, becomes relatively inefficient as scale increases. If such is indeed the underlying production reality, the benefits to be obtained from superior organization are transitory (though still desirable). As national market size grows, the scope for efficient use of labor is diminished even without any increase in its relative price.

A number of issues are raised by the non-homotheticity argument. First, depending on the output threshold at which labor becomes relatively inefficient, many LDCs may continue to derive considerable gains for an extended period if they can produce in a labor-intensive manner. Second, systematic studies of the choice of technique have not, in fact, detected non-homothetic production laws. On the other hand, the technology associated with efficient substitution possibilities at higher output levels may not afford the same level of labor intensity as that exhibited at lower levels. To put these points into a consistent framework it is necessary to employ some simple geometry.

Figure 1 depicts the isoquants perceived by existing producers as consisting of two parts: that to the left of Oz indicates the technical substitution possibilities understood by purely

Figure 1

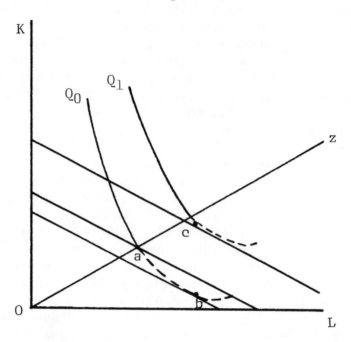

local producers, that to the right the augmentation of factor intensity options made possible by the organizational abilities of MNCs. The cost minimizing production equilibrium for producing Q_0 with the factor prices depicted are a for the local firm and b for the MNC. Thus, the MNCs exhibit lower capital intensity and hence lower unit cost. The isoquants are depicted as being non-homothetic. Thus, as output expands from Q_0 to Q_1 with relative factor prices constant, the minimum cost equilibrium moves from b to c for the MNCs on whose behavior we shall now concentrate. Clearly, if successively higher levels of output were to be produced along these isoquants, growing capital intensity would occur unless the wage-rental ratio were to fall continuously.

However, at higher levels of output it does not appear to be true that the isoquant map remains non-homothetic. Rather, most existing studies, such as those of the Strathclyde group, imply that above some minimum level of output, the isoquants appear to be homothetic, though this has not been

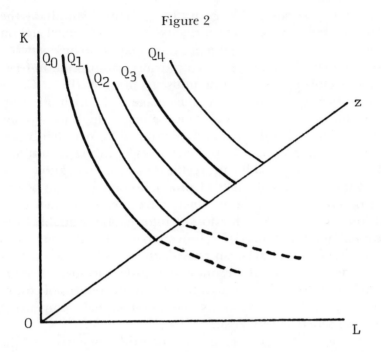

Figure 2

demonstrated in a systematic manner. Figure 2 combines the non-homothetic isoquants characterizing lower output levels with the homothetic isoquants describing higher output levels. To the right of Oz and until Q_0 there are efficient labor-intensive alternatives as shown in Figure 1. Above Q_1 and to the left of Oz the solid isoquants exhibit homotheticity. At output levels of Q_2 or more the expansion path with constant factor prices will be along a straight ray from the origin, unlike the potentially backward curving path associated with a non-homothetic production function.

It must be remembered that just as operation on the dashed segments of Q_0 and Q_1 will require substantial organizational ability to identify substitution opportunities, operation close to Oz in the homothetic part of the isoquant map will require similar skills.

B. Policy Issues

What policy advice emerges from the preceding? First, if the issue of "technology" is going to be defined simply as the DCs making available proprietary industrial knowledge at low or zero cost to the recipients, there may be no link between such knowledge and greater employment absorption in LDC manufacturing. The types of industries in which the LDCs are likely to have a comparative advantage and which offer significant numbers of jobs are not the newer, high technology sectors where production knowledge (apart from the hardware) is of great importance. In product areas as diverse as clothing, food processing, and furniture, the transmission of know-how and the operational tricks of production are likely to be unimportant. Most of the products have been in existence for a long time; the combination of information available from capital goods producers, freely available trade knowledge, and occasional consultants can provide the basis for efficient production. The costs of "technology transfer" in such areas are likely to be low, limited to a fraction of the returns of equipment producers, insofar as the international competition among sellers is very great.

If potential export areas are identified that require the transfer of complete production processes, including detailed instruction on the actual execution of the various stages, these may indeed require high payments. However, more than casual empiricism suggests that such activities are usually quite capital- and skill-intensive and are unlikely to be characterized by a comparative advantage for the typical LDC. It is incumbent upon those who believe that reduced costs of transfer of high technology are a critical issue to demonstrate the appropriateness of the industry choice. Otherwise there is a good chance that, to the existing "urban bias" in development, one may add a high technology bias, both strategies advocated by those who profess to be interested in "the poor" but whose policy proposals somehow always seem to aid those in the upper deciles of the income distribution.

What then are the policies to be followed by the DCs? First, they must be predicated upon a receptive environment in the LDCs. If LDC governments view older technology or used equipment as maintaining a neocolonial or dependent status and thus discourage their use, either by fiat or by failure to confer existing tax incentives on their deployment, there is limited scope for increasing the labor intensity of manufacturing.[28] Similarly, though the advice may be tiresome at this point, it remains true that in the atmosphere of high tariff barriers and quantitative restrictions, businesses have only limited incentives to search for labor-intensive, cost-minimizing techniques. Many other profitable uses of their time are available. All of this being said, and assuming a country wants to improve job opportunities in manufacturing, do the developed countries have anything to contribute?

There has been much discussion, particularly within the U.N. agencies, of technocratic solutions to the problem of providing information on appropriate technology to potential LDC investors. This information dissemination partially replicates the internal information structure of adaptive firms and therefore would be of interest. Presumably, detailed, machine-readable production information would be stored in Geneva or Vienna, and U.N. employees could speedily respond to (letter) inquiries from Karachi or Mombasa after a firm's request for relevant information had worked itself up through the local technology referral service. Such a system, though providing well-paid employment in pleasant middle-European surroundings to a cadre of programmers, engineers, and industrial analysts, must nevertheless be evaluated in terms of its potential benefits to its clients. Would such an international referral service actually alter LDC factor choices? This prospect, to me, seems highly unlikely. Apart from the obvious but nontrivial problems, such as the long leadtimes and the inability of an agency like the UNIDO to demonstrate in a dozen years that, as an institution, it understands the issues comprising optimal choice of technique, the overwhelming objection is that a complex, centralized, bureaucratic mechanism is extremely unlikely to be able to compete in any dimension with private sector salesmen of capital goods.[29] Specificity about performance characteristics, adaptability to raw materials of various quality and to voltage fluctuations, and compatibility of machines of various firms at different stages are all part of the typical salesman's knowledge, as is his ability to answer often unpredictable questions. To take a simple example, a central storage facility would have to indicate the following (selected) characteristics of each loom offered by a large subset of manufacturers from the United States to India for each of the many types of cloth producible:

1. loom speed
2. reed width
3. horsepower
4. cloth width
5. ends of warp
6. picks per inch
7. length on beam
8. weight on pirn
9. input weft per yard of cloth
10. stops per loom running hour
11. mean repair time
12. loom efficiency
13. output per machine hour
14-23. each of ten levels of labor skills required
24. effects on output per hour of yarns of varying quality

The effect of voltage fluctuations, variations in material quality, and skill levels would have to be assessed correctly for each of these characteristics, and the implications of each of these for antecedent and succeeding machines in an entire production process would have to be examined. The combinatorial problems of simply scanning the process make it unlikely that these will be dealt with successfully.[30] While in some sense salesmen face similar problems, they deal with them on a continuing basis and narrow the range of reasonable variation as experience accumulates. While an adequate follow-up system could duplicate this learning, such monitoring and learning has not been a notable feature in other public activities. While the goals of a referral service may be correct, it is unlikely that they are efficiently achievable through such mechanisms, though some well-designed experiments may be worthwhile.

What positive role then can be identified in the technology

sphere for DCs desiring to aid the industrialization-employment objectives of LDCs? Can any DC policies be designed to implant within LDCs those attributes of MNCs that enable them to adapt technology and indeed often improve upon the performance of local enterprises? Clearly the policy decision depends on the identification of those attributes critical to such a performance. Though they are not "provable," my candidates for the critical skills are long production-floor experience (familiarity with earlier methods that work) and a good knowledge of the machinery market. Are there U.S. policies which can effect a transfer or implantation of such skills? Clearly the answer is no if this process is attempted for all countries and all firms; there are simply not sufficient supplies of skilled manpower. However, a major effort in a few countries should be possible and constitute an experimental test of old-style technical aid, newly packaged.

While this is not the context in which to provide the details of such a policy, it should be noted that careful monitoring would be required to determine if the provision of such skills to an LDC would yield any positive effects. Careful selection of firms, with some being given technical aid and some not, would permit such evaluation by local research institutions. A full analysis not only would have to consider whether firms being helped altered their factor proportions, but it would have to include a full cost-benefit calculation of such adaptation. The technical aid envisioned could be handled within the institutional framework of the several technology review authorities already established in countries such as Mexico and the Philippines.

There are other measures as well: decreasing the imperfections in the used equipment market and providing guarantees against "lemons" could improve the employment absorption potential in LDC manufacturing. None of these programs would require huge resource flows, nor do they lend themselves to simple or simplistic policies. They do not conform to the technology demands of proponents of the New International Economic Order, whose link to improved productivity and employment performance is tenuous at best. However, such policies would be more likely to move LDCs in the right direction with respect to factor proportions than would the "free" transfer of advanced technology.

Notes

1. Four recent surveys of the rapidly growing literature on microsubstitution possibilities provide a large number of examples of such findings for specific industrial products as well as in construction activities. These are: S. N. Acharya, "Fiscal/Financial Intervention, Factor Prices and Factor Proportions: A Review of Issues," Staff Working Paper 183, IBRD (1974); David Morawetz, "Employment Implications of Industrialization in Developing Countries: A Survey," *The Economic Journal* (September 1974). Frances Stewart, "Technology and Employment in LDC's," in Edgar O. Edwards, ed., *Employment in Development in Developing Nations* (New York: Columbia University Press, 1974); and Lawrence White, "The Evidence on Appropriate Factor Proportions for Manufacturing in Less Developed Countries: A Survey," *Economic Development and Cultural Change* (October 1978). Acharya's review offers an insightful analysis of the difficulties with industry-wide cross section and time series estimates of the elasticity of substitution and thus provides the rationale for emphasis on evidence from individual process-level microeconomic analysis.

2. The reader interested in the technical details will have to consult the documents from which the data are drawn. Convenient summaries of most studies can be found in *World Development* (October 1977), an issue devoted entirely to recent work at the University of Strathclyde. The technical engineering descriptions of alternate plants are provided in the original, lengthier studies cited in the individual articles. The textile data are from H. Pack, "The Optimality of Used Equipment: Some Computations for the Cotton Textile Industry," *Economic Development and Cultural Change* (January 1978). References to the basic textile engineering data will be found there as well.

3. Other evidence may also be cited. See, for example, the studies in A. Bhalla, ed., *Technology and Employment in Industry* (Geneva: ILO, 1975).

4. For a description and analysis of technical choice in Korea see

Y. Rhee and L. Westphal, "A Microeconometric Investigation of Choice of Technology," *Journal of Development Economics* (September 1977).

5. Of course, the criterion for choosing an optimal sector or project should be social cost-benefit analysis along the lines of Little-Mirrles or Marglin-Sen-Dasgupta. Our implicit assumption is that capital intensive plants would not be internationally competitive when evaluated at social prices.

6. Greater productivity per worker solely attributable to learning will not necessarily increase realized productivity per worker unless several conditions are satisfied.

7. Y. Kawaguchi, "Developing Countries' Exports of Non-electrical Machinery," mimeo, The World Bank, March 1978.

8. D. Lecraw, "Direct Investment by Firms From Less Developed Countries," *Oxford Economic Papers* (November 1977). James Pickett, D.J.C. Forsyth, N. S. McBain, "The Choice of Technology, Economic Efficiency, and Employment in Developing Countries," in Edgar O. Edwards, ed., *Employment in Developing Nations* (New York: Columbia University Press, 1974).

9. See for example, C. Cooper and R. Kaplinsky, *Second Hand Equipment in a Developing Country* (Geneva: International Labour Office, 1974).

10. The exemplar of such work is that of Barnet and Muller, *Global Reach* (New York: Simon and Schuster, 1974).

11. There is no evidence of such effort on the part of LDC-based firms in the capital goods industries. See H. Pack, "The Capital Goods Sector in LDC's," mimeo, The World Bank, 1978.

12. The same conclusion is arrived at by a different route by Stephen Magee, "Information and Multinational Corporation: an Appropriability Theory of Direct Investment," in Bhagwati, ed., *The New International Economic Order* (Cambridge: MIT, 1977).

13. A good summary of the reasons why MNCs face a lower cost of capital than equivalent domestic companies is provided by D.J.C. Forsyth and Robert F. Solomon, "Nationality of Ownership and Choice of Technique in Manufacturing in a Less Developed Country" (*Oxford Economic Papers*, July 1977), p. 260.

14. Ibid.

15. Ibid., p. 278.

16. David J. C. Forsyth and Robert F. Solomon, "Restrictions on Foreign Ownership of Manufacturing in a Less Developed Country: The Case of Ghana," mimeo, 1977.

17. Grant L. Reuber, H. Crookell, M. Emerson, and G. Gallais-Hamonno, *Private Foreign Investment in Development* (Oxford: Clarendon Press, 1973).

18. White, "Appropriate Factor Proportions."

19. The study described is that of G. K. Boon, "Technological Choice in Metalworking, with Special Reference to Mexico," in Bhalla, *Technology and Employment*.

20. Samuel Morley and Gordon Smith, "Limited Search and the Technology Choices of Multinational Firms in Brazil," *Quarterly Journal of Economics* (February 1976); R. Hal Mason, "Some Observations on the Choice of Technology by Multinational Firms in Developing Countries," *Review of Economics and Statistics* (August 1973); William H. Courtney and Danny M. Leipziger, "Multinational Corporations in LDC's: The Choice of Technology," *Oxford Bulletin of Economics and Statistics* (November 1975); Danny M. Leipziger, "Production Characteristics in Foreign Enclave and Domestic Manufacturing: The Case of India," *World Development* (April 1976).

21. Lecraw, "Direct Investment."

22. The author notes these at some length.

23. Samuel A. Morley and Gordon W. Smith, "The Choice of Technology: Multinational Firms in Brazil," mimeo, Program of Development Studies, Rice University, Fall 1974.

24. H. Pack, "Employment and Productivity in Kenyan Manufacturing," *Eastern African Economic Review* (December 1972) and "The Substitution of Labour for Capital in Kenyan Manufacturing," *The Economic Journal* (March 1976).

25. Louis T. Wells, Jr., "The Internationalization of Firms From the Developing Countries," in T. Agmon and C. Kindleberger, eds., *Multinationals from Small Countries* (Cambridge: MIT Press, 1977); and "Foreign Investment From the Third World: The Experience of Hong Kong-based Chinese Firms," mimeo, Harvard Business School, 1977.

26. The failure of the larger firms to export remains to be explained but is a familiar puzzle in other contexts as well. While noncompetitive cost levels are the most obvious explanation, other issues are likely to be involved.

27. See, for example, J. Katz, M. Gutkowski, M. Rodrigues, and G. Goity, "Productividad, Tecnología Y Esfuerzos Locales de Investigación Y Desarrollo," Inter-American Development Bank, March 1978.

28. The best policy would be to eliminate many of the existing investment incentives.

29. Thus, a recent and typical UNIDO publication, *Furniture and Joinery Industries for Developing Countries* (New York, 1977), devotes 239 pages to "processing technology" without any discussion of machinery which would be appropriate in an LDC, unless Finland is included in this category.

30. See Larry E. Westphal, "The Methodology of Investment Planning in the Non-Process Industries," mimeo, The World Bank, 1976.

4
International Transfer of Technology to Developing Countries: Implications for U.S. Policy

Nathaniel Leff

Introduction

With the emergence of technology as a prominent item on the international policy agenda, considerable attention has focused on the obstacles to the transfer of technology to the less developed countries (LDCs). Much of the discussion here has centered on supply conditions, such as monopoly power, which may inhibit the transfer of know-how to the LDCs. Serious problems also exist, however, on the demand side. These obstacles must be identified and analyzed if effective policy measures are to be devised to accelerate the flow of technology for development.

This paper analyzes two central features of LDC demand for advanced technology. We first consider economic aspects of Third World demand, focusing particularly on industrial technology, where LDC concerns have been most intense. In addition, LDC demands in the field of technology have a political dimension. Indeed the political dimension may be more important than the economic, particularly in intergovernmental relations such as the 1979 United Nations Conference on Science and Technology for Development. Accordingly, we will also consider these political demands, which are premised on the assumption by many Third World governments that changes in the present technology transfer system would help to solve important development problems. Such politically motivated pressures cannot realistically be discussed, however, in isolation from the overall economic and policy situation in which the oil-importing LDCs now find themselves. We must therefore consider that broader context, and assess the possibilities for using international measures in technology policy in order to help achieve LDC objectives. These include the aspiration, often voiced by Third World leaders, to diminish their external dependency. This concern for reducing foreign dependency appears as a policy goal or constraint throughout the paper.

This analysis leads to a consideration of the scope and limitations for utilizing technology as a major instrument of U.S. foreign policy in relations with the developing countries. The paper then proposes several specific measures that U.S. policymakers might consider for the coming U.N. Conference. We conclude with a longer term perspective on transfer-of-technology issues in the context of U.S.-LDC relations.

Demand Conditions

LDCs that want to import advanced technology in a specific area usually have available several alternative suppliers. Technical aid, professional education, and equipment salespeople are clearly important sources of technology transfer. In addition, consultant firms that specialize in the design and start-up of new factories are another potential supplier of advanced technology. Finally, multinational corporations (MNCs), both from the United States and from other more developed countries, are another possible source of advanced technology for developing countries.[1] Such companies are available as potential suppliers of know-how either through direct foreign investment or, if the LDC prefers, through licensing arrangements between local and international firms.

Because many alternative sources of advanced technology are usually available as potential suppliers of know-how in specific fields, monopolistic control over the international supply of technology to LDCs rarely exists in practice. A smooth international flow of technology, however, requires favorable demand as well as supply conditions; and here some problems do exist.

For LDCs to import technology effectively, they must be able to identify and articulate their know-how needs. This involves both specifying the objectives they want technology to achieve, and being aware of the possibilities for using different technologies to attain similar goals. Further, the technology importer must be willing and able to search internationally for the best terms available among potential alternative sources of know-how. The first supplier contacted for a specific project may well not offer the conditions that best suit the LDC. This implies the need for expanding contacts among potential know-how suppliers, as well as having knowledgeable LDC negotiators who can evaluate the merits of alternative proposals. These two conditions, effective international search and articulation of technology needs, have not always been satisfied in the developing countries. The "horror stories" sometimes recounted in the transfer of technology area—cases in which LDCs paid exorbitant know-how fees, or purchased technology that manifestly did not suit their needs—usually relate to such instances.

LDC demand conditions may also constrain the international flow of know-how in another important way. As in any demand context, it is essential to distinguish between "needs" and effective demand. In many instances, potential technology importers may be unable or unwilling to support their technology desires with effective purchasing power in the marketplace. We will return to this obvious and fundamental point below. Before doing so, however, we must consider the costs and prices involved in the international market for the transfer of technology, for this is another subject on which misconceptions and recriminations often prevail.

The prices charged in technology transfer agreements typically involve two components. One is payment to the technology-exporting firm to compensate for the direct expenses (e.g., in personnel training) which it incurs in the transfer. In addition, a price is charged for the use of the proprietary technology. The MNC may have "off-the-shelf" know-how available without cost. But the same technology has a positive economic value to the Third World firm, for the LDC company can use the technology to increase its returns by reducing costs or by entering new product lines.

A bargaining situation thus exists to establish the price actually charged for the use of the technology. As in any bargaining context, the price determined in a particular licensing negotiation depends on the specific situation. The more numerous the potential know-how suppliers, the greater the competitive pressures on a particular licensor to reduce the price charged for the technology. But if LDC negotiators are not knowledgeable, or have not searched among alternative suppliers, this advantage may easily be lost. And even under favorable market conditions, the absolute level of the prices charged to technology importers can be high.[2]

One further point is essential for understanding the prices charged in the international transfer of technology. The MNC already possesses the know-how as a result of its past R&D and production experience. At the same time, this know-how has considerable value for an LDC firm, for it can increase its returns once the technology is made available. As in any case involving a good which has value but which has been produced with sunk costs, the price charged reflects economic rents which are determined by negotiation. Notwithstanding its undisputed rhetorical appeal (to both sides), the concept of "fair" or "equitable" price has no economic meaning in this context. A greater awareness of this fact might reduce the mutual hostility and resentment that often pervade LDC and MNC discussions concerning the transfer of technology.

Technology and Development Problems

We have thus far considered the transfer of technology in a narrow economic perspective. This may miss the essence of the story. Current LDC pressures in the technology area are preeminently a political thrust, and partly reflect an expectation that changes in the international availability of technology will help the countries solve some of their development problems. This dimension of LDC demand is especially important in the context of intergovernmental negotiations and the 1979 U.N. Conference on Science and Technology for Development. It is therefore essential to consider the major development problems the LDCs now face, and to assess the extent to which changes in technology, and particularly international policies in this area, can be expected to alleviate the LDCs' social and economic problems. This discussion leads to sobering conclusions. But it is necessary to clarify the realistic possibilities, if only to avoid unwarranted enthusiasm and subsequent disillusion.

Open and disguised unemployment have been increasing in many developing countries, raising questions of widespread economic hardship and political instability. Consequently, attention has focused on the possibility that fundamentally new, "appropriate" technologies might be devised, which would greatly increase employment per unit of output. The possibilities for international policy measures to do much in alleviating this problem, however, are extremely limited.

First, there is little empirical basis for expecting that "appropriate technologies" can in fact be invented which would have a sizable impact on unemployment in the developing countries.[3] Moreover, research and development efforts to devise such technologies will require a gestation period. This is no justification for failing to search for such techniques. But it does indicate that even under favorable conditions, "appropriate technologies" should not be expected to have a short-term impact on unemployment in developing countries.

Further, even if techniques do become available which might utilize much more labor and less capital, they will not be adopted in LDCs unless these countries correct factor-price distortions that now make it profitable for private and state capitalist enterprises to substitute capital for labor.[4] One reason that appropriate technologies have not been supplied to developing countries is that current government policies ensure that they are generally not demanded in the market. The situation is complicated by the fact that LDC intellectuals, policymakers, and technology importers often have very different perceptions of the technology that is appropriate for the country's needs.[5] For technology exporters to become involved in LDC determinations of what is appropriate would, of course, be an unwarranted trespass on local autonomy.

International policies can do little to correct the LDC factor-price distortions that presently inhibit the development and adoption of labor-intensive technologies. The policies that affect the internal relative prices of labor, capital, and foreign exchange are deeply rooted in local economic and political reality. Consequently, external proposals to change LDC wage, monetary, and exchange rate policies can only be an infringement on local sovereignty. And with their current sensitivity to neocolonial domination and dependency, LDC governments can be expected to be especially hostile to such outside interference. Indeed, even well-intentioned international organizations can expect to have little influence in this area. The Employment Missions of the International Labor Organization have had a meager policy response to their recommendations on changing relative factor prices in developing countries.

The theme of "appropriate technology" has also been raised in another sense. Attention has been focused on the development of useful products suited to the needs of the broader population rather than the inappropriate consumption goods purveyed by MNCs and their local imitators. In some instances, this theme has blended into a call for LDCs to reorient their development strategies toward providing the basic needs of the masses in their countries. Again, however, supply can do little to change the situation unless demand patterns also shift. The elites in control of most developing countries have shown little desire to reorient their development patterns toward basic needs. They have also been tenacious in resisting efforts at income redistribution which would provide the effective mass demand for "appropriate" consumption products. In the face of present income distribution and demand patterns in the LDCs, the call for appropriate consumption goods is often little more than a hollow slogan.[6]

International measures in the field of technology can also do little to accelerate rates of output growth in the Third World. Attention here has centered on policies to narrow the international technology gap that now separates producers in the LDCs from producers in the more advanced countries. This focus neglects the fact that the major technology condition holding down productivity in the developing countries is not international differences in technology but rather an *internal* technology gap within individual LDCs.

In most activities, firms do exist in the LDCs which utilize technology similar to that used by producers in the more advanced countries. Most firms in the same sector, however, use techniques that yield markedly lower levels of productivity.[7] The coexistence of high- and low-productivity techniques reflects the weakness of competitive pressures (including competition from imports), or the factor-market imperfections which lead to a dualistic pattern of development. Thus the internal technology gap reflects LDC industrial organization policies, exchange rate regimes, and factor pricing. In these conditions, increased *international* availability of highly productive techniques to Third World countries will do little to increase overall levels of productivity and output growth.

International policy measures can do little to narrow the internal technology gap within individual LDCs. As just noted, the disparity often has its source in LDC policies that distort factor prices and inhibit competition. In agriculture, the slow rates of internal technology diffusion may have other causes, but again outsiders have a minimal role to play. The U.S. government did in fact develop an important policy innovation which could be used to accelerate the spread of modern technology in the agricultural sector of developing countries. This was the Peace Corps. For their own reasons, however, governments in many LDCs have rejected use of this instrument of technology diffusion.

Finally, the potential of technology to raise overall rates of output growth in the developing countries is especially limited because of the special conditions that now constrain economic growth in the vast majority of LDCs—those which import petroleum. As is well known, the countries affected most severely by OPEC's 1974 quintupling of petroleum prices have been the oil-importing LDCs. This is because the sharp rise in oil prices implied a shortage in the foreign exchange available for the other imported inputs necessary to sustain economic expansion in their countries. With high oil prices imposing an import constraint on the economic development of most LDCs, technology is of very limited relevance for their short- and medium-term growth problems. Not only has this new restriction emerged to limit their growth possibilities, but the energy technologies available in the more developed countries offer the LDCs little prospect for a solution to their difficulties.

Technology, and particularly international technology measures, can thus be expected to have little impact in resolving the growth and unemployment problems of the LDCs. What are the prospects for technology policy to aid in another area—reducing external dependency in Third World countries?

Alternative LDC Goals in the Transfer of Technology

LDC efforts in the transfer of technology area aim at three distinct aspects of reducing external dependency. One goal is to relax external constraints on the local supply of goods, by making available to producers within the LDCs the know-how necessary to produce advanced-technology products. This objective involves the *use* of advanced technology for economic development.

Another LDC objective focuses on the possession, rather than simply the use, of modern technology. This objective reflects the concerns of Third World countries to utilize technology as they see fit—reducing external controls on royalty payments and on local decisions involving technology which is supplied from abroad. Similarly, this focus stems from LDC aspirations to be the autonomous masters of the know-how they import. This cannot be achieved if critical elements of technology are controlled by MNC suppliers.

Finally, recent Third World discussions of technology transfer also indicate a third objective. This is the desire of the LDC to increase its capacity to generate technology indigenously. Reliance on technology developed in the advanced countries keeps the LDCs in a state of neocolonial technological dependency. Consequently, many LDCs are now concerned with enhancing their internal technology capabilities by expending a much larger share of their income for domestic

R&D rather than paying for imported know-how. Logically, this objective might appear to reject the whole concept of technology transfer. It has in fact been included in LDC discussions concerning the inadequacies of the present system.

The problem confronting LDC policymakers in the area of technology transfer is that these multiple policy objectives usually involve trade-offs rather than complementary relations: policies that achieve more of one goal may do so only at a cost of lower performance with respect to other objectives. In the present context, LDC policies oriented toward reducing technology payments and MNC control over know-how may also lead to fewer licensing agreements being negotiated. Similarly, the lower the cost and the greater the volume of technology transfers, the less Third World countries will be able to generate the capacity to develop industrial technology domestically. For with an elastic supply of technology from overseas, firms in LDCs are unlikely to incur the resource costs and uncertainties involved in developing know-how locally. Indeed, the most effective policy for developing domestic R&D capacity would be drastic curtailment of international technology transfer, for that would divert local demand for technology to potential domestic suppliers. Such a policy, however, would entail heavy costs in terms of relaxing technological constraints on local production of high-technology products.

These trade-offs and complexities are often very much part of the context in which LDC governments must formulate their policies on the transfer of technology. The existence of multiple and competitive objectives also helps explain why LDC policies in this area may often act at cross-purposes, and thus suboptimize with respect to all three of their policy goals. Finally, this situation also has important implications for U.S. foreign policy.

Technology Transfer and U.S. Foreign Policy

Because LDC objectives concerning the transfer of technology often suffer from inherent inconsistencies, U.S. efforts to satisfy LDC aspirations will be beset by the same contradictions. Thus, if U.S. policy is oriented to be responsive to the needs of the LDCs, it will inevitably "fail" in one or another of the policy goals just discussed. For example, measures to lower the cost and accelerate the flow of international know-how will conflict with LDC desires to develop indigenous technological capacity. If this outcome appears remote, recall the effects of U.S. efforts to provide a large volume of cheap food imports to the LDCs. Such measures were castigated for inhibiting the agricultural development of the LDCs. The most effective U.S. policy for promoting enhanced LDC R&D capabilities would be a U.S. embargo on the export of technology, a measure which would shift LDC demand for know-how from foreign to local suppliers. Such a drastic policy would, however, conflict with Third World aspirations for increasing production of industrial products and domestic mastery of advanced know-how. Similarly, U.S. policies to increase LDC possession of imported technology may inhibit the flow of new know-how, and thus retard LDC production of high-technology products. The United States is thus in a "no-win" situation: regardless of its policy on the transfer of technology, it will not be able to still LDC critics on this issue.

Entering into dialogue with the LDCs on the transfer of technology is also unlikely to yield positive results for another reason. Fruitful dialogue is feasible when the participants share common premises. This condition is not satisfied in the transfer of technology issue. Many Third World leaders consider know-how a free good, the common heritage of all humankind. This is not the position of American firms who possess proprietary technology. And even if the U.S. government were to accede to the LDC view, it lacks the means to make good on such a stance by controlling the actions of private corporations in this area. In addition, LDCs have come to believe that changes in the international availability of technology will have a dramatic impact on their development. As we have seen, however, as long as the internal LDC technology gap persists, increased international transfer of know-how will do little to accelerate Third World economic growth. Under these conditions, efforts to intensify U.S.-LDC dialogue on the transfer of technology will not be productive. And far from blunting political pressures, more discussion will lead only to exacerbated dissension, as disagreement on basic principles becomes increasingly clear. Similarly, a U.S. negotiating approach which appears to validate LDC expectations concerning the results to be anticipated through intergovernmental negotiations would also have unhappy consequences. False hopes can lead only to disillusionment and frustration.

Finally, forceful American policy in this area is in any case not likely to be feasible because of the difficulties the U.S government faces with respect to its own policy objectives and instruments. These limitations are particularly evident in the field of industrial technology, which has been the special focus of LDC pressures. Transfer of technology has, of course, been opposed by organized labor in the United States. Many of the changes contemplated by the LDCs—replacement of direct investment by transfers of technology to nonaffiliated firms; lower royalty fees; and an end to confidentiality requirements—are resisted by much of American business. Moreover, general public opinion may also oppose changes which appear to imply government acquiescence to a decline in America's technological superiority.

These domestic political conditions clearly constrain the capacity of the U.S. government to formulate a policy objective consistent with LDC desires. Moreover, most industrial technology is in the hands of American corporations. The U.S. government lacks the means to control technology transfer decisions by these firms. The problems the government faces within the United States in this field are clear if we contrast the transfer of technology with, say, the shipment of agricultural surpluses to the LDCs, an issue-area where domestic interests made feasible the implementation of an effective U.S. policy.

These considerations suggest that transfer of technology is not a promising issue for the United States to cultivate in its relations with the developing countries. The U.S. economy clearly has a strong comparative advantage in technology. But the same does not apply to the U.S. government's capacity to use industrial know-how as a foreign policy tool within the context now set by LDC aspirations. And U.S.-LDC differences in assumptions and in interests are too wide on this issue to dissipate in well-intentioned dialogue.

In order not to damage relations with the developing countries, then, U.S. foreign policy would do well to downplay the transfer-of-technology issue. This is all the more necessary now, to head off proposals that would place technology at the very center of U.S.-LDC relations in the 1980s. The likely consequence of that approach would be false hopes in the LDCs, and subsequently, resentment of the U.S. government for its failure to deliver. A refocusing away from the technology issue may make the United States appear unresponsive to the international policy agenda which is presently proposed by the developing countries. But as a major power, the United States can also take an active role in determining that agenda. And the United States is within its rights to avoid having the agenda give prominence to items that will further impair U.S. relations with the Third World.

U.S. Policies to Accelerate Technology Transfer

Although transfer of technology is not an issue that our government should seek to emphasize in its relations with the LDCs, the United States will undoubtedly be pressed to make some policy moves in this area. A list of specific proposals follows:

1. A major constraint on the flow of technology to the developing countries has been the capacity and willingness of the LDCs to pay for imports of know-how. Consequently, the U.S. government might consider instituting a policy whereby an LDC government would identify specific technology needs, and the Agency for International Development (AID) would request bids from American companies to supply the requisite know-how. AID would then pay the costs of the ensuing program for consulting, training, licensing, and management fees. This new policy would be especially relevant for the poorest LDCs, on which U.S. aid is now focusing. It could also be used, however, to help other developing countries translate their technology needs into effective market demand.

Technology Transfer: Implications for U.S. Policy

The new policy proposed would be an obvious instance of using "private interests for public purposes." The policy would constitute a major extension to the private sector of "contracting out" programs, such as AID has long utilized with universities and other not-for-profit institutions. This extension is logical in the transfer of technology context because most product-specific industrial technology is the proprietary knowledge of private firms.

The proposed policy would offer American firms the clear advantage of an expanded market for their know-how. It would also facilitate the unbundling process in activities where this is economic, and hasten the transformation of American firms from sellers of goods to sellers of know-how in the developing countries. The policy might thus be especially beneficial from the viewpoint of overall U.S. relations with Third World countries. By increasing the attractiveness, both to firms and to LDC governments, of an alternative to direct foreign investment, it would reduce the intrusive presence of MNCs in the developing countries. Finally, a further extension of the policy might entail having the U.S. government purchase and own specific technologies as a public good. These technologies could then be made freely available to LDCs as subsequent requests appear.

2. The greater the stock of new technology which American firms possess, the greater its flow to the LDCs is likely to be. Consequently, additional U.S. subsidies for domestic research and development by American firms would also be beneficial. In particular, vastly expanded R&D efforts which would increase the flow of new technology would accelerate the depreciation of existing know-how. With their technology increasingly transformed into a wasting asset, American firms would be less reluctant to sell know-how to developing countries. And with American technological superiority maintained with new know-how, opposition in the United States by labor and the general public to export of know-how would diminish. Changes in the tax treatment now accorded U.S. R&D expenditures would probably be the most effective means for implementing such subsidies.

3. In addition to greatly expanded support for domestic R&D, two other tax changes are also relevant. Section 861 of

the Internal Revenue Code disallows as a business expense the portion of R&D costs that is allocated to overseas sales. At the same time, the LDC authorities also refuse expensing for R&D expenditures incurred in the United States. The ensuing higher tax payments lower the returns to American firms which produce advanced technology products in the Third World. Such local production, however, can be an important source of technical progress in LDCs. Consequently, a change in Section 861 may well be justified.

Similarly, Section 911 of the Internal Revenue Code increases the tax liabilities of Americans who are engaged in technology transfer activities overseas. Translated into higher salary costs for U.S. firms, this condition reduces the incentives for technology transfer, or increases the costs to LDC technology importers.

4. The transaction costs involved in negotiating inter-firm agreements for the license and transfer of technology are largely fixed costs which do not vary with the scale of the sale. Consequently, small- and medium-sized firms that are unfamiliar with overseas technology demand may face an indivisibility which hampers their entry to the market in exporting know-how to Third World countries. This limitation on technology transfer is unfortunate, for such firms often face internal institutional constraints that would otherwise incline them to "go the licensing route" rather than engage in direct investment in order to increase overseas returns on their know-how. Consequently, a U.S. government program to subsidize the entry costs for small- and medium-sized firms which are potential technology exporters might be helpful. The Department of Commerce program for nascent exporters of goods may be a useful model here.

5. Outside of the area of industrial technology, much of the know-how desired by the LDCs cannot be supplied by American firms, for they do not have the technology to transfer. This limitation applies particularly in the area of "basic needs" with technology which is relevant and appropriate to LDC environments. Consequently, the U.S. government might create, jointly with LDC governments, technology centers to do research and development work on such know-how needs.

The energy, food, housing, and health areas might offer a high potential return to such efforts.

Established on a bilateral basis and located within individual LDCs to increase their effectiveness, such technology centers would constitute a source of know-how which is not now available. Such centers would also answer to the LDC call for enhancing Third World R&D capabilities. Staffed by local personnel, they would create a demand for LDC technologists, and might thus mitigate problems of brain drain.

Creation of such technology centers would be an attractive symbolic gesture for the American government. Forethought and control would have to be exercised, however, to avoid having them replicate the disappointing experience of another institution created in an international gesture, the Vienna International Institute for Applied Systems Analysis. If the U.S. government is interested in having these research centers generate helpful technology results, it will have to confront potential difficulties which often recur in such projects. These problems include the lack of a well-defined research focus, inadequate responsiveness to the needs and conditions of potential technology users, and poor diffusion of research results. The difficulties are likely to apply with special severity in Third World countries because of well-known conditions related to the sociology of science in the LDCs.[8]

6. The 1979 Conference on Science and Technology for Development may well propose the establishment of a new U.N. agency for technology transfer. Such an agency might usefully centralize information on the technologies available in specific areas, and on the costs of transfer arrangements with alternative MNC suppliers. By reducing search and negotiation costs for technology importers, the agency would clearly serve LDC interests. It would be especially helpful for the poorer LDCs, which lack experience on alternative suppliers and terms in the international market for know-how.

The interest of the United States in the creation of such an agency is less clear. It might well be opposed by American companies that export technology. It would also involve an additional bureaucratic layer, which may delay technology transfers. In view of these diverse perspectives, it would be well to prepare in advance the U.S. government's position concerning the creation of such an agency. A relevant condition here is the fact that the United Nations could establish an agency for technology transfer without U.S. assent.

7. Finally, two policy recommendations which are often advanced in this context should *not* be considered without extreme caution. One is the suggestion that the United States should greatly increase its aid for the education of scientific and technical personnel in the Third World. Indiscriminate growth on the supply side without concomitant expansion in the local economic demand for such technological infrastructure will not produce the desired result of enhanced local scientific activity. As the experience of countries like Argentina, Jamaica, and Chile indicates, the outcome is likely to be scientific underemployment, frustration, and brain drain. Say's Law does not operate here: an increase in the supply of scientific personnel does not generate a parallel increase on the demand side.

Second, the proposal is sometimes heard that the United States should act to raise the quality of existing institutions for scientific and technological education in the LDCs. The problem here is that well-intentioned efforts by outsiders to upgrade Third World universities often conflict with local sensitivity to intrusiveness and dependency. Thus proposals to modernize university curricula are often resented as an infringement on the autonomy of local faculty. Efforts to raise educational standards by imposing heavier laboratory assignments or tougher examinations are understandably resisted by student militants. Political opposition can also be mobilized because of the elitist and antidemocratic character of such changes. To avoid aggravation of hostility to the United States, the task of improving LDC educational institutions might best be left to local people.

A Broader Perspective

U.S. policy can gain effectiveness in the area of transfer of technology by viewing the issue in its broader political and economic context.

During the postwar period, most developing countries have achieved high rates of economic development. They have also made enormous technological progress in terms of the new products they produce, and the techniques they use to produce them. This experience suggests that LDCs have in fact been able to overcome technological constraints on economic growth and structural change in their countries. This is also indicated by the fact that when the pace of economic development has slackened in individual LDCs, the causes have usually been aggregate balance of payments problems, or internal political crises rather than inadequacies in the international transfer of technology. The mechanisms for this successful technological development have varied: importation of equipment; bilateral and multilateral technical aid; international consultancy activities; internal education programs; international research on high productivity agriculture; direct foreign investment; and licensing agreements with multinational corporations. These sources of advanced technology have provided know-how directly. They have also made possible the production of new goods within the LDCs, and the ensuing accumulation of local technological experience through learning-by-doing.

Notwithstanding the extraordinary growth of their technological capacity during the last thirty years, the LDCs are increasingly unhappy on the issue of technology. This is clearly a case where aspirations have increased as a result of achievements. Moreover, the more developed countries have also experienced rapid technical progress in the postwar period. Consequently, the LDCs contemplate an international technology gap which has continued undiminished despite their achievements. The fact that LDC frustration has mounted in the face of this situation is hardly surprising. This experience does indicate, however, that the international technology problem is likely to persist (or worsen) almost regardless of what the more developed countries do in this area.

The Third World's intensifying focus on the technology issue also stems from the current economic and policy predicament of its member nations. As discussed above, the pace of economic development in most developing countries is now limited by the oil-price import constraint. International policy measures, however, now offer little hope to the LDCs concerned that they will escape the prospect of an impending downturn in their rates of economic development. For some years, LDCs entertained hopes that implementation of the much-discussed New International Economic Order would save the day for them. A U.N.-sponsored study by W. Leontief, however, made clear how large the increase in aid and import expansion of the more developed countries would have to be to significantly narrow international income differentials.[9] In the absence of political will in the more developed countries to implement such measures, the U.N. study can indeed be read as a *reductio ad absurdum* of the New Order. LDC governments have in fact now become much more realistic in their expectations about achieving a New International Economic Order.

Facing poor economic prospects and with little likelihood that other items on the international policy agenda will yield them positive results, the developing countries have turned to the magical "black box" of technology. As noted earlier, however, there is little basis for expecting that technology can do much to accelerate short- and medium-term growth for most developing countries. This limitation is especially apparent in the face of the foreign exchange constraint on development which is now imposed by OPEC's pricing policy. Nevertheless, the technology issue is not likely to go away, for it has deep roots in the LDCs' economic and policy predicament.

Two implications follow from this discussion. It would be helpful if American policymakers would devise measures outside the technology area which would answer to the needs of the developing countries. The formulation of such an alternative policy agenda is beyond the scope of the present chapter, but two prime items can be mentioned. These are the maintenance of vigorous economic expansion and of open markets in the United States—conditions which would facilitate the growth of Third World foreign exchange receipts from exports and from capital inflow. Continuing focus on the technology issue, however, will lead to a bitter learning process for LDC policymakers, as they come to appreciate the limited scope for international measures in this area to solve their economic problems. In the interim, U.S.–Third World relations will probably worsen.

Learning will also proceed among two other important groups of actors in the transfer-of-technology process—MNC know-how exporters and LDC technology importers. The LDCs are currently attempting to reduce the licensing rates and shorten the contract periods of the know-how agreements through which they import technology. Their efforts here are facilitated by the entry of many new technology sources from Japan and Western Europe, and the ensuing intensification of competitive pressures on the supply side. In the future, LDC negotiators will increasingly probe the international technology market in an effort to develop new terms and new suppliers.

The international market for technology is thus undergoing major changes. Importers and exporters can learn the contours of the new market only by experimentation and evaluation of subsequent experience. Thus American MNCs in some industries may try to maintain previous contract provisions, and attempt to enforce their will by withdrawing from the market if their terms are not met. The extent to which they can succeed with such an approach is not yet apparent. Similarly, it remains to be seen whether the LDCs can attain their price and contract objectives without reducing the volume and quality of international technology flows.

As in any learning process, the adjustment to changing conditions may involve pain to the parties involved. Consequently, one can anticipate that continuing outcry will be a permanent feature of the transfer-of-technology arena. The ensuing call for policy action is likely to keep technology transfer on national and international policy agendas.

Thus, both because of the LDCs' economic and policy predicament and because of the changing conditions in the international know-how market, issues related to the transfer of technology to developing countries cannot be expected to recede after the 1979 U.N. conference. This is unfortunate in view of the conditions, discussed earlier, that make technology an especially poor area for American foreign policy. Nevertheless, because of the need for longer-term analysis and policies, it is all the more important that the U.S. government develop an analytical and institutional framework for dealing with transfer-of-technology issues on a continuing basis. This is obvious. It bears mentioning only because of the understandable tendency to neglect such longer-term perspectives in the face of the short-term pressures for the 1979 conference.

Conclusions

Although attention has often focused on possible supply constraints, demand conditions also limit the transfer of technology to developing countries. A smooth flow of know-how requires that LDCs be able to specify their technology needs and that they search internationally for the most appropriate source of supply. Effective demand also presupposes that developing countries are willing and able to pay for the know-how they desire to import. In most technology fields, multiple sources of advanced know-how exist internationally. Accordingly, competitive pressures operate to reduce the prices charged for the importation of technology. Nevertheless, the absolute level of prices can be high; hence LDC capacity to pay for know-how is a major determinant of the volume of international technology flows.

Despite problems in this area, the developing countries have achieved rapid progress in industrial technology in the past thirty years. In agriculture, progress has been slower. Here, however, the major problems have not been in the international transfer of know-how. Agricultural technology is largely specific to particular locations.[10] Consequently, what is required is the development and diffusion of new techniques suited to the special environmental and economic conditions in the developing countries. Diffusion of the higher productivity techniques already available is especially important. But there is little that outsiders can do in this area.

LDC pressures in the area of technology also have an important political dimension. As we have seen, however, LDC demands here are often inherently inconsistent. Moreover, international measures in technology transfer can do little to increase employment, accelerate economic growth, or reduce external dependency in the developing countries. In addition, the U.S. government is likely to have serious problems of its own in formulating and carrying out an effective policy for transfer of industrial technology to the Third World. I have proposed some specific measures that U.S. policy makers might consider in preparation for the U.N. Conference on Science and Technology for Development. For the reasons discussed, however, transfer of technology does not appear to be a fruitful area on which to focus future U.S. relations with the LDCs.

It is encouraging to recognize that, notwithstanding the problems on which attention currently centers, technological progress in the developing countries will probably accelerate in coming decades. The present system for the international transfer of technology functions far more effectively than is commonly assumed. Moreover, a key input for the growth of technological capacity is the stock of experience obtained through the production of goods with modern know-how. This stock has been accumulating rapidly in most developing countries. The investments made in LDC agricultural research should also yield important innovations, for that research is now conducted on an increasingly large scale. Finally, the large sums the LDCs have devoted to expanding their educational systems during the past two decades should also bear fruit in higher rates of internal technological diffusion, both in agriculture and in industrial technologies. Consequently, if one is concerned with LDC technological progress per se, rather than with the broader economic and policy issues which are raised in connection with the coming U.N. conference, there is, in fact, considerable room for optimism.

Notes

1. For a formal model, see Ronald Findlay, "Relative Backwardness, Direct Foreign Investment, and The Transfer of Technology: A Simple Dynamic Model," *Quarterly Journal of Economics* (February 1978).

2. Donald J. Teece, *The Multinational Corporation and the Resource Cost of International Technology Transfer* (Cambridge, Mass.: Ballinger, 1977).

3. Richard Eckaus, *Appropriate Technologies for Developing Countries* (Washington, D.C.: National Academy of Sciences, 1977).

4. Lawrence J. White, "The Evidence on Appropriate Factor Properties for Manufacturing in Less-Developed Countries: A Survey," *Economic Development and Cultural Change* (October 1978).

5. See, for example, Simon Tietel, "On the Concept of Appropriate Technology for Less Industrialized Countries," *Technological Forecasting and Social Change* (April 11, 1978).

6. Denis Goulet, *The Uncertain Promise: Value Conflicts in Technology Transfer* (New York: IDOC/North America, 1977), Chapters 6 and 7.

7. See, e.g., Joel Bergsman, "Commercial Policy, Allocative, and X-Efficiency," *Quarterly Journal of Economics* (August 1974), pp. 409-433.

8. See, for example, the papers by R. Ramasubban and by Stephen Hill in S. S. Blume, ed., *Perspectives in the Sociology of Science* (New York: Wiley, 1977).

9. Wassily Leontief et al., *The Future of the World Economy*, (New York: Oxford University Press, 1977).

10. Robert Evenson, "International Diffusion of Agrarian Technology," *Journal of Economic History* XXXIV (March 1974).

5
Technology Transfer in Practice: The Role of the Multinational Corporation

Samuel M. Rosenblatt
Timothy W. Stanley

This chapter is based on a series of interviews with officials from half a dozen major multinational corporations involved in international activities that include some form of technology transfer in both developed and developing countries. The purpose of these interviews was to evaluate the findings and conclusions of the other research in this study against the specific experiences of representative multinational corporations in developing countries. (Summaries of these individual discussions are contained in the Annex to this chapter.) These discussions generally involved two or more individuals with extensive corporate experience in each company's business operations in developing countries. In some instances those interviewed had firsthand, on-site operational factory experience in a developing country. In other cases the contacts were more familiar with the company's overall developing country operations and technology relationships and less with production-line activities. The direction of the discussions that took place with the individual companies naturally reflected these differences.

These interviews were not an attempt to develop general conclusions that could readily have widespread application for U.S. public or private policy in the developing countries. With such a small sample, no attempts at quantification were considered. The information elicited was totally dependent on the memory of the persons being interviewed, and no site visits were made to developing countries to observe production lines or other plant operations.[1] The broad point of view expressed was, of course, that of the multinational corporation, although it was colored by individual outlook and experience. These discussions indicate the general nature and intensity of involvement with developing countries of six reasonably diverse and representative multinational corporations, and they reveal the shifting perspectives on this involvement brought about by changes in circumstance and environment. The attitudes of the participants toward the roles of their own company, the U.S. government, and the developing country private and public sectors in the mutually beneficial uses of technology owned or provided by multinational corporations in developing countries also come across in these discussions. It is hoped that they will provide relevant additional evidence to be considered in the framing of the U.S. position for the United Nations Conference on Science and Technology for Development (UNCSTD).

The topics in these interviews ranged from the global aspects of relations between industrial and developing countries to the general business relations between a company and a developing nation, to the specific aspects of the company's technology transfer and technology adaptation practices in individual technology transactions. In short, the discussions included the broadest international policy considerations as well as the details of the company's interaction with developing countries and with affiliates or other local businesses in those countries. Although relatively little of a "proprietary" nature was discussed in the meetings, they were conducted with the specific understanding that the names of participants and their companies would not be disclosed, in order to encourage maximum frankness in the discussions. It was later suggested, at the institute workshop, that permission be obtained to identify the specific companies, inasmuch as proprietary or confidential information was rarely discussed. However, this was not feasible because of the delays that would have been involved in the necessary company and participant review of the summaries.

Research Findings Evaluated

As noted above, a major purpose of the interviews was to

compare the findings of leading academic researchers with the practical experience of companies involved in technology transfer and thereby gain additional insights. The preliminary findings of the authors of the preceding three chapters that were drawn upon in the company interviews are summarized below.

By way of background to his analysis in Chapter 2, Gustav Ranis begins by noting that not all appropriate technology is labor-intensive and not all appropriate goods are so-called basic goods. Most important, he places the science and technology policy issues for developing and industrial countries within the broad contexts necessary for a full understanding of the possibilities and limitations of the role that science and technology might play in assisting the economic advancement of the developing countries. While there is considerable that technology and technology transfer have done, and much that they will continue to contribute, their potential must be viewed in a practical perspective. Furthermore, the constraints imposed on these potential contributions by extraneous or otherwise conflicting developing country social and economic policy objectives must be made explicit to be fully understood.

There appears to be a wide range of technological alternatives among production processes that are subject to local adaptation possibilities from which developing countries may choose, with a somewhat lesser range of alternatives regarding appropriate goods. A large number of choices actually may continue to be inappropriate. The reasons for this pattern can be found on both the demand and supply sides of the market. On the demand side, distorted factor prices and the absence of significant competitive pressures on producers are foremost among these reasons. Ranis argues that the developing countries should look toward the development of export markets as a way of opening their domestic economies to the forces of competition, rather than pursuing the more limited policy objectives of import substitution. Similarly, more balanced growth between rural and urban areas would help generate a more effective demand for appropriate goods, which in turn could stimulate the type of technological innovation that would be conducive to the adoption of appropriate technological processes.

On the supply side, Ranis cites the high search costs associated with finding appropriate processes; some limitations imposed by patent restraints, apart from the difficulties of trying to value separately the technology component from the rest of a foreign investment package; the inefficiency of information dissemination within most developing countries; and the failure to integrate the science and research communities in developing countries into an institutional and incentive framework that supports the practical needs of the nation's producing community. External supply constraints, such as those imposed by some multinational firms and developed country governments, have also acted to circumscribe the supply of alternative technologies.

In attempting to develop suitable policies to cope with these conditions, Ranis recognizes the vital role that has been played by local adaptations and incremental modifications to technology at the working level of the firm or plant, and he argues that this role should be stressed in the future. He also refers to some needed modifications in macroeconomic demand policies—such as establishing workable competition, eliminating factor-price distortions, and emphasizing rural development—that would help bring about more effective demand. On the supply side, he mentions such possibilities as improved information flows, modified patent policy (perhaps to include some unbundling of technology where feasible), and the development of research and development centers of excellence in developing countries for a limited number of carefully selected industries. He concludes on an optimistic note about what can be accomplished in this area, by both the developed and developing countries, though the major burden must, of necessity, fall on the latter group. They also have the major portion of the opportunities and responsibilities for bringing about these required changes.

From his survey of literature and his own analysis, Howard Pack concludes that multinational corporations have tended to be quite adaptive in their choice of techniques for developing countries. This is borne out, he notes, by comparisons with locally-owned firms in developing countries and with the techniques used by the multinational in its home advanced country market, and even with the scale of operations that would be suited to the smaller markets in developing countries. There are, of course, limitations to this adaptiveness and developing countries in particular should avoid trying to induce the establishment and growth of clearly inappropriate, advanced-technology, heavily capital-intensive industries. Again, the cost of acquiring information about alternatives must be factored into the analysis and, for certain producers, because of the high opportunity cost associated with gaining knowledge of alternative techniques, it may be perfectly rational to select a less labor-intensive mode of production. Pack notes, parenthetically, that this cost is likely to be a lesser restraint on a multinational corporation, due to its established worldwide network of operations, than for an indigenous firm.

In developing his policy recommendations, Pack emphasizes the importance of the right industrial mix for the developing country to pursue, in conformance with whatever comparative advantage these countries may enjoy. Such an approach is the most likely to offer significant employment opportunities. This set of industries is also unlikely to consist of high-technology operations that are research-intensive and of recent vintage. He also stresses that, whatever policies are pursued, they "must be predicated upon a receptive environment in the LDCs."[2] By a receptive environment he means the absence of dogmatic opposition by the developing country to the use of older technology or used equipment, as well as avoidance of an "atmosphere of high tariff barriers and quantitative restrictions"[3] which can only deter the incentives businessmen have to search for labor-intensive, cost-minimizing techniques. Pack also has little use for any centralized, computerized information system that would spew out the latest data on alternative technology possibilities. Such an approach would inevitably be too cumbersome for the purpose and highly inefficient as well.

In keeping with his analysis, Pack would encourage policies for the developed countries that are "designed to implant within LDCs those attributes of MNCs that enable them to adapt technology and indeed often improve upon the performance of local enterprises."[4] Those attributes would focus on practical skills acquired through long years of operation on the production floor and an intimate knowledge of the machinery market. The industrial countries might seek to implement this policy for a few selected industries in a limited number of countries. Again, as with Ranis, this approach is not spectacular, nor does it conform to the highly ambitious but unrealistic aspirations that some developing countries have been voicing in their policy statements on technology as part of their call for a New International Economic Order.

Nathaniel Leff's approach and analysis differ from those used by Ranis or Pack. He examines the aggregative, or macroeconomic, nature of the demand for technology, and concludes that the potentials for new initiatives for the United States and other industrial countries in this area are highly constrained. What must be done must be undertaken primarily by the developing countries themselves and the industrial countries must neither appear to nor actually intervene in the internal affairs of developing countries in order to get them to adopt a particular set of policies.

Leff reviews the technology transfer transaction between the multinational corporation and a developing country. He contends that the bargain struck is generally the result of hard negotiations between two sides that are reasonably equal in market knowledge, where neither the buyer nor the seller has an enduring monopoly advantage. He also sees little realistic prospect that appropriate technology, under prevailing conditions in developing countries, can play a significant role in the development policy of these countries.

This conclusion is based, in part, on the existence of factor-price distortions in these countries and on the fact that little incentive or motivation exists for developing countries to change this situation. He finds the whole question of technology transfer policy rife with ambiguities, on both the developed and developing country sides, and therefore considers it unlikely to lead to fruitful results. He would hope, rather, that both sides would downplay technology and its transfer—whether as a negative force that has restrained growth or as a "black box" that will open up the secret to growth—and concentrate on policy questions that are more likely to bring mutual benefits.

Summary of Interviews

Tables 1 through 3 summarize the general comments, reactions, and recommendations that resulted from the six company interviews. The tables are organized to move from the narrow to the general, drawing on the key issues outlined above. Table 1 focuses on the content and form of the technology transfer agreement itself; Table 2 addresses the economic aspects of the business arrangements that the multinational corporations have established in the developing countries, and Table 3 looks at the broader aspects of the multinational corporations' relations with those nations.

The following points, which are in general agreement with the previously summarized research, are the most significant to emerge from these discussions:

1. The multinational corporations have a clear preference for working directly with their private sector counterparts in the developing countries. Moreover, the multinational corporations generally continue to have good working relationships with these companies once they are established. These relationships have developed best when there was a minimum amount of government involvement and the two private companies were free to deal directly with each other. The U.S. multinationals generally sought to deal with persons or firms about whom they had reliable knowledge.

2. Not surprisingly, the multinational corporations are concerned about what they perceive to be a deteriorating investment environment in many developing countries. A persistence or increase in this negative attitude might result in a redirection of investment flows and of some of their other international activities away from the developing countries toward other industrial countries. This unfavorable atmosphere, which engenders increased hostility and higher political risk, has caused some shifting by the multinational corporations in the risk-reward ratios they use to make business decisions to the detriment of the developing countries.

TABLE 1
Summary of Nature of Agreement

Company	Licensing	Duration of Agreement	Fee Structure	Equity Position	Capital Repatriation	Company to Company	Long-Term Relationship	Unbundling
A	Technical assistance program involves close working relationship	—	Percentage of sales with minimum fee structure used occasionally	Prefers equity position	No major problem indicated	Prefers this relationship with limited LDC government involvement	Looks toward long-term relationship for mutual benefit	Not a major concern
B	Uses service contracts	—	Cost plus fixed fee	None	Limited problem	Works with LDC government or government agency	Looks toward long-term relationship with LDC	Not a major concern but will tailor offer to changing LDC demand
C	All transfers in form of licenses	10-year initial agreement preferred	Percentage of sales basis	Prefers equity position and provides more support in such instances	No major problem indicated	Works with wholly or partially owned affiliates	Looks toward long-term relationship as basis of mutual benefit	Not a major concern
D	No licensing, only investment	Long-term investment	—	Prefers 100% ownership, will take majority but tries to avoid minority interest	Sensitive to need to anticipate difficulties in this area	Works only with wholly or partially owned affiliates	Looks toward long-term relationship	Not a major concern
E	All transfers generally in form of licenses	10-year initial agreement	Percentage of sales basis	Varies—has 100% equity but also deals with government agencies	Aware of increasing problem in this regard	—	Looks toward long-term relationship	Willing to work in any mode; with unbundling value of technology per se more difficult to establish
F	Works primarily through investments with occasional licensing	—	—	Prefers 100% ownership, will take majority interest but tries to avoid minority interest	No major problem indicated	Works closely with private companies in carrying out investment decisions	Looks toward long-term relationship	Not enthusiastic about expansion of this approach

TABLE 2
Summary of Business Operations

Company	Labor Intensive	Training	Vintage Technology	Business Operations	Skill Level	Repair & Maintenance	LDC Firm Competition	Advanced Country Competition
A	Limited possibilities in peripheral operations	Tied to technical assistance agreement	Limited—more prevalent in South to South transfers	Technology transfer relates to all aspects of business operations	Middle level skills are major bottleneck	—	Limited—and not strenuous	Limited
B	Limited possibilities in peripheral operations—constrained by technology of product	Extensive and highly involved in candidate selection	Possible—due to technological constraints on related facilities	Technology transfer relates to all aspects of business operations	Retaining lower skill levels a problem	Encourages LDC initiatives	—	Fairly active competition
C	Limited by safety needs—results in white collar inefficiency	Fairly active, person to person is best	Limited due to absence of market incentive and knowledgeable manpower	More involved when it has an equity position	Plant operations most critical	—	Limited	Extensive, encompasses overall operation
D	—	Extensive at all levels of operation	Older equipment already in use in LDCs—some reluctance on LDC part to upgrade	Technology transfer relates to all aspects of business operations	No particular constraints noted	Lacking in the LDC so included in plant operations	Limited	Extensive
E	Tailors product to market	Tied to technical assistance agreement, mixed results	Willing to provide older technology especially in response to LDC requests	Limited to specific license agreements	No particular constraints noted	—	Very limited	Varies with LDC
F	Attempts to adjust production processes used in core operations to reflect markets and resources of developing country	Extensive with emphasis on use of developing country instructors	Used in both developed and developing countries according to market to be served	Heavily involved in accordance with degree of investment	No particular constraints noted	Built into internal operations	Very limited	Intense

TABLE 3
Summary of Broader Implications of Agreements

Company	Relations with LDCs	LDC Environment	Reverse Flow	South/South	Regional Research Centers	General Comments
A	Prefers minimum involvement by LDC government	—	Limited transfer of technology in this manner	Occasional transfer of technology in this manner	Conceptually sound but practically difficult	Mutual trust between companies essential. Establishing real value of technology that will be acceptable to LDC. Recognition by LDCs of their own priorities.
B	Generally favorable	Political instability and turnover of governments	None	Limited	Conceptually sound but practically difficult	Recognizes need of receptive attitude by host LDC.
C	—	Receptive LDC environment crucial; MNCs may choose to invest elsewhere	—	Conceptually an attractive possibility	Expense may be too great	Establishing real value of technology that will be acceptable to LDC. U.S. antitrust practices constrain transfers.
D	Sensitive issue especially in centrally planned LDCs	Worsening LDC environment may drive out MNC investment	—	—	Very limited in reality, especially for higher skills. At lower skill levels practical problems of implementation.	Unanticipated business problems are likely to occur. Excessive LDC restrictions. Recognition by LDCs of their own priorities.
E	—	—	Limited but possible	—	Selected individual LDCs may be capable of developing their own R&D capability	Need to differentiate among LDCs in terms of technological sophistication. Establishing real value of technology that will be acceptable to LDC. Recognition by LDCs of their own priorities.
F	Preference for export but developing country objectives necessitate investment	Generally hard-nosed bargaining by developing country	None	Possibility of larger technology flow in future	Absence not a constraint	Developing countries should have policies supportive of their own comparative advantage. Need to pursue less protective tariff policy to encourage foreign investment

3. Within the constraints imposed by the economic requirement to earn a competitive rate of return on their resources committed in developing countries, the multinational corporations are supportive of the economic development aspirations of the developing countries. Their realization would not only benefit the citizens of developing countries, but would, of course, provide some expanded economic opportunities for the multinational corporations.

4. The effective transfer of technology to developing countries involves an all-encompassing effort. It requires a continuing involvement by the multinational, working closely with the personnel of the recipient company, whether this receiver be a wholly-owned subsidiary, a participant in a joint venture, or an independent business. The forms of support that the multinational corporation provides can extend beyond the legal contract that exists between the two entities. Furthermore, the multinational must generally anticipate business problems that would not normally arise at home. A key element that the multinational tries to impart is the managerial competence to cope with these situations. Technology, in fact, is very much integrated with managerial skills at the shop or plant level.

Nature of Agreement

As noted, Table 1 summarizes the involvement of the individual multinational corporation with developing countries and shows that the companies used both direct investment and licensing to transfer technology. Under either of these approaches, the multinationals have found themselves involved in all aspects of the business operations, well beyond their initial expectations. In licensing situations, the multinational corporations generally viewed a ten-year initial agreement as the minimum period required to establish a successful relationship. The earlier years of an agreement were seen pretty much as learning opportunities for both the multinational and the developing country. Participating-country insistence on shorter periods could discourage the initiation of these agreements and the long-term business relationship which is a key factor in corporate decision making about overseas activities.

The fee structure in licensing and other technical assistance arrangements is generally based on a percentage of revenue, as an incentive to both parties to make the relationship an active one. This structure evolved in reaction to claims that some multinationals were content to pursue their arrangements with developing countries in an unaggressive manner—in effect, to sit on their technology and not exploit it.[5]

The multinationals much prefer to establish an equity position with their developing country counterparts, even when they only license technology or provide technical assistance. Indeed, in the case of investment the multinationals would opt for 100 percent equity, if possible, or at least majority ownership. Their principal reason for seeking this arrangement is the greater control it gives them over the actual operations of the developing country entity. However, in view of the changing preferences among developing countries in this regard, these corporations were also willing to be flexible and follow the established pattern of the country or industry. Whatever the legal form of the association, the multinationals generally sought to establish long-term relationships in the developing country. As is the case with most sound business operations, the corporation spokesmen recognized that efforts to make quick large profits would only prove counterproductive. (Although the attitude of this small sample is not necessarily universal, it is generally shared among a large majority of the companies known to the authors of this chapter. A "get-in and get-out" attitude on the part of multinationals is, in their experience, a rare exception.)

Two points emerged from these discussions regarding multinational corporation views on "unbundling" and capital repatriation that were somewhat surprising because of the insistent position that developing countries have taken on them. First, most of this group of multinationals did not view with alarm the reported desire of the developing countries to unbundle the technology package. In some instances, it should be noted, the potential for unbundling did not exist because the multinational provided only the equivalent of a service or consultative transfer. However, even when a fuller range of services was offered, the multinational companies interviewed did not appear to be unduly troubled by the prospect of separating parts of the package. However, they did express some concern about their ability to convey to the developing country a realistic appreciation of the full cost of this unbundled technology. Since the essence of the technology would be contained in a more abstract form, and frequently tied to management processes as opposed to being embodied in a product, the historical costs associated with its development would be both less apparent and perhaps less appreciated. Thus, while unbundling, per se, did not appear to be an obstacle, establishing the basis for a mutual agreement on the true value, or cost, of a technology process has been and will continue to be a major sticking point in multinational corporation relations with developing countries.[6]

The second surprising point was that capital repatriation was also viewed by these companies as less of a stumbling block than had been anticipated. Those interviewed were certainly aware of a recent general tightening by the developing countries on this issue, but to date they had not encountered too much difficulty with it. Historically, developing country complaints about excessive charges for imported technology and products may have been associated with multinational corporation attempts to circumvent the restrictions developing countries had placed on capital repatriation. The complaints may also have reflected isolated instances in the past when shoddy merchandise or equipment was delivered to the developing country by a foreign firm in quest of a quick and one-time high rate of return. All the companies interviewed regretted such past practices (which they associated with less well-established and less responsible concerns) and tended to view their recurrence as highly unlikely. Such behavior, in any event, was also seen as inconsistent with any interests the multinational corporations might have in establishing long-term relationships with the developing countries.

Business Operations

Table 2 summarizes those aspects of these multinational

corporations' involvement in developing countries that directly relate to their business transactions and production operations. All but one indicated that they were able to achieve only a limited adaptation of advanced country production processes and techniques within the context of developing country resource endowments. This finding is not surprising, given the structure of the interviews, the small sample, and the limitations previously cited on research not verified in the field. Such adaptation as these five corporations were able to accomplish generally took place in the so-called peripheral areas of production, such as materials handling, storage, and warehousing. With the exception noted below, little effort was made by the companies to tailor the production process or the existing machinery, expressly to the resource mix of the country. In some instances, the constraining factor was the technology itself; in others it was concern for safety and health.

The one exception to this general response and reaction occurred in a company whose operations were normally heavily capital-intensive. It did greatly modify its production processes to meet the operating conditions it found in developing countries, especially the comparatively high price there of capital relative to labor. The cost of capital goods was pushed up by the highly protective tariff policies pursued by some of these countries. Another factor that influenced this decision was the limited size of the local market to be serviced by the plant. In any event, because of these comparative prices, the company was forced to develop special inserts for use in its production line; these modifications not only added to the labor costs of production but also created a demand for a reasonably high level of skilled labor. In addition, the company slowed down the operating speed of some of its tooling equipment, because of the expense of obtaining such machinery. This had the effect of doubling the useful life of the equipment and was thought worthwhile even though the slowdown reduced the efficiency of the equipment somewhat.

This was an instance where nominal costs of resources more or less reflected their true costs as determined by the relative supply and demand for these resources in these countries. In many developing countries the opposite is true, because their governments have distorted prices and lowered the cost of capital in an attempt to encourage its greater use, believing this to be the most expeditious route to economic development. While this presumption generally no longer holds in most developing countries, the practices it created are still prevalent.

All of the multinational corporations were heavily involved in operating training programs for the local populace, which of course was in their own self-interest since they needed trained people to operate their plants successfully. However, their commitment to these programs appeared to go beyond their own needs and revealed their belief that training is the backbone for the overall development of the economies and industrial infrastructures of these countries. All the companies interviewed recognized the necessity to train persons at all skill levels, including the development of local senior managers and supervisors. They also were aware of the need to limit the number of their own or other industrial country nationals who function in a host developing country.[7]

However, two of the companies made it a deliberate practice to use teams from one developing country to go to another and train its nationals in a start-up operation. Training and skill development enhances the potential mobility of labor and thereby contributes to the overall development of the developing nation's economy; however, it also creates the possibility of a loss of skilled manpower throughout the country if an effective demand for this type of labor is not generated and maintained by the developing country. Three companies cited instances where particular skill levels were a possible bottleneck, but no two of them named the same skill.

There was a surprising mélange of responses on the use of vintage, or older, technology. Some multinational corporations saw only a limited potential for its use in developing countries—because the talent needed to recreate it and make it operational no longer exists or because there is no market incentive to do so. Moreover, any U.S. government subsidy to provide such an incentive was viewed unfavorably because it would likely result only in an inefficient use of resources. Nevertheless, other companies saw a considerable potential for the older technology. In one instance, in fact, owing to the limitations of related equipment, only the older technology was suited to the developing country environment. In another case, the older technology was already in use and the attempt by the multinational corporation to upgrade this equipment was not viewed favorably by the developing country itself because of a reluctance to expend its limited foreign exchange. Another multinational used the older equipment in both developing and developed country economies in accordance with the market to be served. Energy efficiency may prove to be an increasingly important criterion in future technology application.

As noted above, the multinationals generally found themselves supporting the full range of activities normally associated with running a successful business, whether located in a developed or a developing country. Their involvement with a developing country business entity, which is more extensive if the multinational corporation has an equity interest, extends to such functions as plant design and construction, accounting and financial management, and marketing and distribution services. In effect, the multinational acts as a management consultant for all phases of the operations of the developing country company.

The U.S. multinationals interviewed claim that competition from their counterparts in other industrial countries is intensifying, as more and more firms prove competent to provide the types of services and functions they perform. Moreover, the developing countries are quite capable of fostering this rivalry when it suits their purposes. However, the competition from indigenous developing country firms in the product market where the multinational corporation is already established has not been too intense, in the experience of those interviewed.

Broader Implications of Agreements

One of the outgrowths of their involvement is the support that has developed among these multinational corporations for the economic development aspirations of the developing countries. Again, while there is a certain self-serving aspect to this attitude in terms of potential markets and investment outlets, most of the multinational corporation spokesmen who participated in these discussions are also interested in seeing these countries succeed in their development plans. (A number of them had substantial overseas service and, hence, exposure to the problems of developing countries.) Many of them indicated a sense of frustration over the developing countries' inability to do a better job of ordering their own economic development priorities, assessing their own resources and needs, and establishing an ordered time frame for satisfying these needs. A few of these corporation officials, in effect, appeared to be calling upon the developing countries to establish a more planned economy, in contrast to the market-oriented economy they clearly favor for the United States! It was suggested that U.S. government foreign assistance might be used to help improve the implementation of these plans.

Conversely, and in spite of this sympathetic attitude, there was no denying that these same company representatives were concerned about the deteriorating attitude in many developing countries toward the active participation of foreign businesses. This deterioration, so far as the multinationals are concerned, results from a combination of political rhetoric, government-stimulated nationalism, and increasing restrictions on the actual operations of the multinational corporations and on their latitude to take corrective actions where business conditions warrant. The view was expressed that a continuation of these trends could very well result in a scaling down of the activities of multinationals in developing countries, and even in the eventual departure of many of them.

On this point it would be useful to recall Goulet's observation cited in Chapter 1. As one who is sympathetic to the plight of the developing countries, he notes that "The harsh truth is that poor countries do need technology, and there exist few alternative sources outside the TNC's where they may obtain it."[8] It is clear that the developing countries have little to gain by forcing the multinationals out. It is equally clear that the corporations are not interested in confrontations with these countries, which would be fraught with uncertainty and with potentially great losses.

On some related, broader concerns, the corporation spokesmen saw only very limited prospects for a reverse flow of technology from the developing to the industrialized world. One instance cited was a technological adaptation moving from a fairly advanced developing country to a developed country for further implementation. There was more expectation that technology, especially appropriate technology, would move from the advanced developing countries to less developed ones. In part this was due to the existence in these potential sending nations of a vintage of technology that would be particularly suitable to the emerging needs of poorer receiving countries. Such technology already may have been adapted to a developing country environment, so that its further adaptation may be easier to accomplish. Finally, the view was generally expressed that the potential development of regional research and development centers in developing countries was conceptually sound but often impractical. The impracticality stemmed from the limited manpower available to staff and operate such facilities, their large costs, and the rivalries among developing countries that have handicapped such efforts in the past.

Conclusions and Recommendations

What role have the multinational corporations played in the practical aspects of technology policy and technology transfer to developing countries? They have clearly been involved at the heart of the process. Their involvement stems from their expectation of a rate of return on the use of their resources that is competitive with the available alternatives. As a consequence of this involvement they have also acquired a greater awareness of the difficulties that impede the efforts of the developing countries to achieve their economic development goals. This has sensitized them to the practical obstacles and opportunities that exist for the use of technology in these countries. It would seem that they have acted upon the actual opportunities presented to them through their ownership of technology without excessively exaggerating its potential, although such judgments are necessarily subjective. Perhaps a summary that

Lawrence White used in his assessment of the role of multinationals in appropriate technology could be applied in this instance. "Although the MNC's may not be the heroes of appropriate technology, they appear to be far from the villains that many make them out to be. They have the management expertise, and they are frequently willing to use it to adapt to labor intensive processes."[9]

Even if this evaluation is a fair one, the question remains whether multinational corporations cannot make a greater contribution and, if so, what approaches might be useful. The views obtained in the interviews confirm the general findings of the three preceding chapters that technology is too deeply imbedded in the conduct of transnational business, both from a management standpoint and with regard to local conditions, infrastructures, and factor prices to be easily separable or independently emphasized for policy reasons.[10]

Technology's contributions to development are likely to be maximized as market forces and business climates improve, and to be minimized as the converse applies. Moreover, there is relatively little that technology, per se, can do to influence basic economic and political forces. This "bottom line"—which comes from this study and from much of the other recent literature on the subject—may be realistic, but it is not likely to be received as good news in the context of UNCSTD, given the hopes and frustrations of many of the developing country participants. Nevertheless, acceptance of basic realities—including the proposition developed in several parts of this research that the main responsibility necessarily lies with the host countries themselves—may be a prerequisite to achieving any real enhancement of technology's role.

On the other hand, there are some possible policy initiatives, as discussed in the preceding chapters, which the United States and other industrial countries could take and which would be helpful both substantively and politically. Notwithstanding the doubts expressed in a number of our interviews about the practicality of regional technology centers, there may be merit in further experimenting with such centers in one or more regions of Latin America, Africa, the Indian subcontinent, and Southeast Asia. Such centers could have a two-fold mission:

identification of needs and applications of appropriate technology for developing country industry and *simplification* of the search process for the more appropriate technologies available to improve productivity and employment. These centers could be funded under existing multilateral aid programs such as those of the Inter-American Development Bank, the Organization of American States, the Asian Development Bank, or the United Nations. Alternatively, one or more industrial countries could establish a center as a cooperative project under bilateral aid programs. To expedite action, the United States could take the initiative at UNCSTD by offering to fund such a center unilaterally in, say, Latin America for an experimental five-year period if agreement could be reached on its location and terms of reference. A few leading companies in technologically-oriented fields might be persuaded to assign experts or scientists for a year or two to such a center or to a national agency as a public service, or to contribute support in other ways.[11]

In general, companies develop technological innovation as an integral part of their own production and marketing operations. Nevertheless, there are some cases where technology itself is an end product; more could undoubtedly be done under various multilateral or bilateral aid programs to facilitate credits for the acquisition of such technologies, either directly by countries receiving aid or through the regional centers noted above. The key to the success of these efforts would be a sharper identification of specific needs and realistic plans for the application of the technology acquired by the developing countries.

Another approach is the voluntary contribution of technological skills to development needs, particularly in appropriate technology applicable to elementary nutritional, housing, and health programs, as well as rural transportation and communication. Many executives of multinational companies, concerned at the deteriorating political climate they encounter in developing host countries, have wondered what they can do to improve the situation. One answer might be to apply the company's managerial and technological skills to the development of an appropriate technology—for example, solar cooking devices or simple modular housing techniques suited to particular local conditions. This would usually be outside the firm's main line of business but would serve as a voluntary contribution to its community relations effort where the company's particular technological skills were especially applicable. For example, a producer of optical glass might undertake a solar cooker project.

Awards and prizes for particular technological applications which helped meet basic human needs could be sponsored by industrial and developing country governments, professional societies, or chambers of commerce and might serve to stimulate imaginative efforts. Again, one must be realistic about the expectations; but given the attitudes of many international business executives that they are not in the business of technology for development or even of technological transfer, except as part of normal production operations, such consciousness-raising might produce valuable dividends. At any rate, the costs of the attempt would not be large.

It might also be worthwhile to explore the possibilities of a more fruitful three-way partnership involving industrial countries, the more advanced developing countries, and the less developed ones. The middle group of countries has already passed through many of the more critical stages of structural change associated with long-term economic development, yet they are close enough to the poorer developing countries—both in geography and in their own development process—to retain a perceptive insight into the potential rewards and disappointments of choosing various paths of technology. Such a partnership could be built around some of the other institutional approaches mentioned above. The officials of the multinational firms interviewed who had used this approach in their own operations found it to be of benefit to all parties concerned.

Finally, though the authors of this chapter believe that technological adaptation for development could profit by imaginative efforts to capture more of the voluntary and goodwill potential of the business community, it need not rest on such motivations alone. Both home and host country governments might consider tax credits or other incentives for corporate efforts devoted to technological innovation and application to development. This possibility seems especially applicable to the United States, whose structure of business taxation lends itself to specific incentives. But neither goodwill nor incentives can succeed in the absence of the basic disciplines associated with the presence of competitive economic forces in developing countries. Host countries should seek in their policies to utilize such forces to create conditions in which development prospers, and not adopt policies which inhibit them. In the long run, economic forces themselves may prove to be the best "conductor" of technology for development.

Notes

1. Information collected in this fashion has some major limitations. On this point see Chapter 3 and Lawrence T. White, "Appropriate Factor Proportions for Manufacturing in Less Developed Countries: A Survey of the Evidence," Agency for International Development, *Proposal for a Program in Appropriate Technology,* House Committee on International Relations, Committee Print, 95th Congress, 1st Session, Feb. 7, 1977, p. 143.

2. Chapter 3, p. 78.

3. Ibid., p. 79.

4. Ibid, p. 81.

5. One other way of countering this developing country attitude, which did not come up in these discussions, would be to relate the fee structure on patents and licenses or other forms of technical assistance to the profit performance of the recipient business. This would certainly enhance the incentives of the supplying multinational corporation to see the project succeed. A variant of this approach is recommended in Walter A. Chudson and Louis T. Wells, Jr., "The Acquisition of Technology From Multinational Corporations by Developing Countries" (New York: United Nations, 1974), p. 3.

6. Chudson and Wells have the following to say on this point: "The cost of the individual elements [of a technology package] is generally impossible to isolate completely from the total payments by the local affiliate to the parent enterprise. The foreign firm is interested in maximizing the total receipt of funds from all affiliates, whether from dividends, royalties, interest, management

Technology Transfer in Practice 131

fees, profits on raw materials or components sold to affiliates or from other sources. Within limits (including tax regulations) the amount of a particular flow will generally reflect considerations of management control, tax, exchange control, or other factors, rather than the particular element of technology associated with the name of the flow." P. 29.

7. Although quantitative data were not gathered in these interviews, there is other evidence that the number of Americans employed by U.S. multinationals abroad is relatively small, mostly in managerial, finance, and sales positions. A survey by the International Economic Policy Association showed that its respondents' U.S. employees abroad aggregated less than one half of one percent of the total combined employment in their 631 overseas facilities. See Statement of Timothy W. Stanley before the House Committee on Ways and Means on "Tax Reform and Foreign Source Income," March 8, 1978.

8. Denis Goulet, *The Uncertain Promise* (New York: IDOC/ North America, 1977), p. 69.

9. White, "Appropriate Factor Proportions," p. 146.

10. In the mid-1960s many Europeans were preoccupied by a feared "technology gap" with America. Studies and conferences at that time soon revealed that there was really only a "management" gap— which has since narrowed very substantially as European industry modernized.

11. However, it is worth recalling Leff's previous observation that innovative U.S. efforts to accelerate the spread of technology, e.g., the Peace Corps in agriculture, were not always accepted by the LDC governments.

ANNEX TO CHAPTER 5: CASE STUDIES

The interviews involved in these cases were conducted off-the-record in order to assure frankness in the discussions. Identities of the companies and of the corporate officials interviewed have therefore been protected. The six were selected, however, to reflect a broad diversity in lines of business, types of technologies, and geographical areas of operation.

Company A

Background

Company A is a long-established firm with over $100 million in net income. This discussion mainly involved its packaging line of business. It is essentially engaged in transferring process technology to recipient countries. It is active in both developed and developing countries. While it has some hardware for sale, it tends not to treat this as a major element of its technology transfer program. The company apparently acts basically as a mangement consultant to the recipients of its technology and provides process know-how as well as skills in other management areas in order to support the successful operation of the recipient company.

Company to Company

Technology is transferred from a private company to another company, and one government cannot give privately owned technology to another government.

A critical judgment an LDC company must make is selecting its partner on a particular project, because a feeling of mutual trust must develop if the project is to succeed. The decision makers in the LDCs may not know anything about manufacturing the particular product, but most of the people that Company A ends up with as partners tend to be fairly sophisticated in financial affairs. They are also able to make intelligent choices among alternative suppliers of technology. However, in a particular technical assistance project, the company believes the recipients have grossly underestimated the cost of the whole production process, not just the technical assistance part. The company feels they have also overestimated the likely return on investment and is trying to be candid with them on this issue, even though it may eventually lose the contract.

The MNC is looking to a long-term, normalized commercial relationship with the host company and it is not interested in short-term maximization of profits.

Some host companies come to believe that they can get along without any technology support, and after a five- or ten-year period they cancel the contract. What they find is that very shortly they are no longer able to keep up, that technology is a dynamic force, and that improvement and upgrading is a continuous process. Consequently, an LDC company will soon find itself unable to succeed in its own market and will have to look elsewhere for some technology support. Most LDCs and individual companies cannot afford to carry on their own research and development as a replacement to foreign technology. Its desire for a continuing relationship is one of the reasons the MNC likes to establish an equity position in the host country company.

Company A tries to limit its contacts in the host country to its private sector counterpart. It has found that since government representatives are neither technically qualified nor profit-oriented, their objectives vary greatly from its own, and indeed from the people in the LDC who are investing with them. But the company has to work with the government in connection with whatever constraints the latter may impose. In some sense, the vagueness of many laws in Latin America serves as a blessing, since it enables governments there to interpret legislation according to existing needs. For example, the Andean Pact, that started out as a very rigid force that would confine the operations of foreign companies, is in fact being interpreted somewhat more loosely as the needs of the individual member countries require.

Capital/Labor Mix

Less developed countries are on a broad learning curve, so their costs are going to differ from those of developed countries.

Generally speaking, their labor costs per worker per hour are less, but their total cost of production is greater than in a more developed country. And while they are in the learning process, many of these countries will recognize that it is to their advantage to use labor where a developed country would use more mechanization and automation. Where an industrial society would use forklift trucks for warehousing, the LDC would man-carry. There is much less automation in the handling of raw materials and of products. For example, in certain Latin American countries, forty people would work in a particular process that would require only a handful of people in the United States or Europe to push buttons and watch a TV screen. The licensee decides how to bring out a product and determines the mix between capital and labor.

Different technological approaches may be used by various organizations in a particular country. Everybody who makes the same product in a Latin American country uses the same basic type of equipment. It involves a long-established basic design. Company A had refined the applications of this equipment to the point where it was particularly suited to the needs of one country. The result was a highly efficient operation. Some of the other companies in this country also had technical assistance, including one which had a spasmodic relationship with a supplier of technology, but presumably the other companies were not as efficient as Company A's affiliate. Yet if the government were not controlling prices, the host country's affiliate would be able to run everybody else out of the market because its costs were lower than those of its competitors. When the government imposes a minimum selling price, usually at the lowest common denominator of efficiency, any firms above this level are virtually guaranteed profit. By such pricing behavior, the government appears to be subsidizing inefficiency.

The presence or absence of competition is a strong force in determining the use of efficient technology. For example, the fact that one fairly advanced industrial country experiences very little competition was one reason it was behind in technology. The same thing was true in another industrial country.

Technology Transfer in Practice

As regards relative prices and the possibility of price distortions, the best one can do is to live with them as they exist. The local people are well aware of price distortions and their costs to the economy. But they also know there is very little they can do about it, and to stay in business, a company must live with the realities of these distortions.

In the preparation of technical assistance contracts for certain unsophisticated countries, a firm must go back to the basics and show them how to make the product, and send in the relevant machine operators and other lower level technicians. In more developed countries one would send engineers, but poor countries are not yet ready for that advanced level of technical communication.

Export Markets

The marketing radius for the principal product discussed in these interviews is about 200 miles in the United States or 300 miles elsewhere. Consequently, the motivation of Company A in setting up a program in a developing country is usually not the intent to export to a third country—although host country policies sometimes make such exports a criterion for approval.

Technology as Investment

One of the principal problems of the MNCs is to help the LDCs or their representatives understand that the transfer of technology must provide a return on an investment that includes very costly research and development and engineering by the multinational corporation.

Since nearly all of Company A's technical assistance packages are based on a percentage of sales, and not on a fixed price, the company has an interest in seeing the affiliate apply the technology and increase sales as quickly as possible. However, the company may also have to impose a minimum fee and license agreement for two reasons: to help cover the costs of a smaller operation and to guarantee an incentive for the recipient to go ahead with the technology. There is also an unavoidable time lag between the initiation of the technology transfer and the time when the results begin to show up in improved production, sales, and profits.

All recipients essentially receive the same level of technical assistance and all the company's home facilities are open for LDC representatives to observe the current processes and to take back with them any useful knowledge gained. If they buy new equipment, the company will send a man to help set it up. There are also manuals, interpreters, and a number of international seminars on specific technical subjects which the recipient companies are free to attend.

Reverse Flow

In this specific product line, there is very little reverse flow of technology, particularly from the LDCs.

Unbundling

The LDCs are free to choose any of the equipment that is available, not only from Company A but from all other suppliers of hardgoods. None of the contracts that this company concludes has compulsory purchase requirements. Threats by LDCs to unbundle technology would have little impact for the company, since no effort is made to bundle in any event.

Vintage Technology

The recipient certainly has the option of using equipment that was available in the home country twenty or thirty years ago. Within that choice, the recipient can decide how automated he wants to become. If he wants to substitute people for a conveyor, or for a forklift, or for stacking boxes instead of palletizing, he may do so. This may take him back ten or twenty years in terms of technology, but that is his choice. Reservations on the results of such a choice were expressed, however. For example, a product developed with the methods of twenty years ago will probably be comparably behind in quality. Clearly, the methodology will not be geared toward energy conservation and it will not embody the latest quality controls or safety features. Also, it may not be ethical to transfer that type of product—for example, from a safety point of view—even if it were steps or years ahead of the quality then being produced in the LDC. While ethical questions could be raised about any product or the production process it entails, when a more suitable and higher quality product or process is available in the United States, the MNC gears its technology to satisfy the market conditions in a particular country. In one Latin American country, the MNC shipped manufacturing equipment that was obsolete by U.S. standards because a particular aspect of the resulting product absolutely suited the requirements of the LDC market, and the use of the latest processes in that market would have resulted in considerable waste.

It was also pointed out that there are rules in many recipient countries that prohibit the importation of anything except new equipment. To use older vintage technology, in many instances, it would almost have to be reinvented, due to the present unavailability of equipment expertise and processes. (One of Chapter 5's coauthors had an analogous governmental experience with military technology in the early 1960s. Faced with a local insurgency, an allied developing country requested urgent U.S. assistance in enhancing its undeveloped air power. But everything then available consisted of high speed jet fighters, beyond the capacity of the local pilots to operate or their mechanics to service. Moreover, their high performance was a handicap rather than an asset to the particular mission involved. The eventual solution was to buy some World War II fighters from a military museum in a neighboring country and to contract with some ex-pilots and mechanics from that war to put them into service. They did the job and at a total cost well below that of purchasing a single "modern technology" aircraft!)

South to South

Company A has arranged for people from one of the more advanced developing countries to go to other LDCs to assist in the transfer of technology through the use of affiliates and not through the use of home country MNC personnel. In undertaking these transfers, the Company A affiliate uses technology and techniques that may no longer be available in the United States, but are appropriate for the recipient. In the process of training the potential recipient, they help him develop his own capabilities.

Process Technology

Company A is involved in process technology that is constantly renewing itself. The company does not have the obsolescence factor experienced, for example, in the electronics industry where, after a big scientific breakthrough, everything prior becomes obsolete.

Technology transfer really should not be confined to machines or techniques but should include the overall concept of management and the administration of a firm, including the bookkeeping system and even personnel.

Middle Management

The greatest bottleneck to long-term efficient production seems to occur at the middle management level—the technicians, supervisors, and foremen. These are a new class in the developing countries. Usually the top management is very knowledgeable and the pool of unskilled labor is satisfactory, as most of the workers are highly motivated people who aspire to a better life. The lack of depth and training in middle management seems to be the most troublesome area. For this reason Company A has undertaken a great number of training programs to expand the quality and quantity of middle management. The spokesman said, "You don't transfer technology at either the lowest level or at the top managerial level. The real transfer takes place at the middle management level." This transfer is accomplished in a variety of ways, including manuals, exchange visits, training programs, and other avenues.

LDC Priorities

One of the major difficulties that LDCs face, whether they are seeking economic development or articulating reasonable political objectives, is the absence of a broadly-based educational system. A seedbed must be developed first and not enough attention is being paid to that task. One LDC, for example, has misused its resources through expenditures for massive industrial projects and products. But it has ignored its own educational system, which is abominably poor. The time needed to improve this educational base is generally not compatible with some of the political demands for fast results. For example, in one LDC, policy virtually required every department supervisor in a plant to be a mechanical engineer, but the man who reported to him and who really got things done on the plant floor in integrating the technology being transferred could not read or write.

Regional Research

Regional research centers, while ideal in concept, require a great deal of practical compromise and cooperation which may be very difficult to achieve.

Company B

Introduction

Company B operates in the service area, primarily in transportation. The aspect of its operations discussed with regard to technology transfer involved servicing contracts through which the firm provides management operations, maintenance, and training assistance to a foreign government corporation or government entity in the air transportation area. These included such items as support and maintenance of an international airport—including runways, the electrical and lighting system, water, commissary, passenger services, cargo handling, ground equipment, terminal maintenance, navigational operations, radar, and similar responsibilities.

Training

Because of the demanding nature of its operations the company tries to maintain a veto over the selection of trainees and to specify the type of training they will receive. Otherwise, the company fears that nepotism or other forms of favoritism might lead to the employment of unqualified people. In the more safety-sensitive areas the company trains to U.S. standards; that is, to the point of FAA license or its equivalent.

While there are formal training programs as well as person-to-person contact, the most efficient method of transmitting technology has been the latter.

Use of Indigenous Workers

The company tries to use the indigenous population as much as possible. Its approach is to train local employees and gradually pull its own people out as the locals are able to assume increased responsibility. The ultimate objective is to reduce the foreign personnel to as few as possible; to perhaps just one or two serving in an advisory capacity to the senior local official responsible for the overall operations of the facility.

The company has encountered problems across the board in attracting and retaining high, middle, and lower level skilled workers. In terms of retention, the company indicated some difficulty in meeting competitive wages available elsewhere, due to the reluctance of the national airline concerned to raise wages and thereby increase the cost of its operations. The company has therefore lost some talent, including accountants and licensed mechanics, representing a significant investment in training. Many took employment elsewhere in the country or emigrated. The mobility of labor is often less at the higher skill levels, such as pilots, since alternative opportunities are more limited than for the lesser skills.

Skill Mix of Available Labor Supply

For a highly skilled operation such as piloting, the company has generally found a nucleus of people from which to select suitable candidates. The selection process is worked through the government. In one instance the company narrowed a universe of 1,000 people to the twenty or so candidates who were finally selected.

Maintenance and Repair

The company tries to encourage the transfer of responsibility for maintenance of peripheral ground equipment and automobiles to local people as soon as possible. Sometimes this turnover is accomplished too precipitously. In one instance the locals discovered that they really lacked the backup maintenance required, equipment was breaking down, and other problems were developing.

Labor-Intensive Operations

The company has recognized the necessity of providing more labor-intensive operations when it could. For example, it has installed a reservation system that is one step behind the latest computerized version in the United States. This satisfies the desire to employ more labor and suits a market that would not justify the major expense involved in a more sophisticated operation. In some of the peripheral activities such as ramp operations, passenger contact, or baggage handling a more labor-intensive technique will be applied in the LDC, sometimes employing twice the number of workers that would provide the service in an industrial country. However, the technology of the industry does impose a restraint on how far this procedure can be carried. For example, in the newer and bigger airplanes, all the baggage is containerized, so that the LDC airport must have special equipment.

Vintage Technology

Because of technical limitations, airports in some of the developing countries are unable to accommodate modern jet airplanes. Consequently, an older plane is used that would not be suitable to more advanced countries.

Initiative, Renewal, and Long-Term Relationships

The company generally awaits a host country invitation to submit a proposal to assist in setting up its airline and related operations. Once contracts are established, most countries are willing to make renewals and maintain the association. However, some countries occasionally think they can continue without support, mainly for reasons associated with cost reductions. The company is willing to renegotiate its contract at such times by reducing the scope of its services and eliminating some of its support personnel. The company never tries to force itself on a country where it is not wanted. The success of the project depends on a genuine willingness on the part of the host country to seek the cooperation and support of the company. Otherwise the effort will fail, whether or not there is a contract.

Various host countries negotiate very hard on the price of the servicing contract. The company is aware of some international competition for the services it is willing to perform, from U.S. companies as well as foreign producers or airlines. The company generally negotiates a cost-plus fixed fee contract with the national government. There has been some discussion of using a percentage of profits but this approach has not been attempted to date.

Head-to-Head Competition

Generally speaking, the company does not expect to train and develop a competitor; rather it hopes to develop a service complementary to its own. However, it did find that over a long period it had succeeded in training one developing country's airline to the point where it competed directly with the company in other areas. The company believed that, had it not undertaken this program, another competitor would have. However, the prospect of direct competition has never been a major factor in technology transfer for this industry.

Program Benefits

The company has earned profits and created jobs for its own employees through these programs. It has also provided employment and an upgrading of skills to citizens of the recipient country and helped to create a national asset. The operation of air services has also led to purchases of equipment from other U.S. firms.

The company feels that its greatest success has been in the Middle East, in countries which are especially anxious to select qualified candidates and have them trained properly. Cooperation with these countries has generally been very good. The foreign exchange and other financial constraints have of course also been much less of a problem there.

Infrastructure Support

The company has recognized that sometimes the isolated

location of an airport makes it difficult to attract and retain competent people. To deal with this situation, the company on occasion has undertaken to provide some schooling for the children of employees. However, the company recognizes that this is only a stopgap attempt and not a complete substitute for more elaborate facilities provided by the host government.

U.S. Government Role

The U.S. government has been very supportive and encouraging of the company's efforts to develop programs of assistance abroad. The company is sensitive about its international reputation and attempts to do a highly credible job in all countries. This is evidenced by its careful selection of both foreign nationals and local citizens to work in these countries.

Political Obstacles

A major difficulty the company has had in working with the LDCs has been cultural obstacles or political interference that makes it difficult to operate an airline or related services efficiently. In some cases, repeated changes in the government have created problems for the company, since it must demonstrate to each new incumbent that it is doing a credible job. Related to this is the situation where many LDCs that do not have an economic need for a national airline have opted for one for political or prestige purposes, or have oriented their air services to limited international markets rather than to domestic development of the country.

South-South Transfer

This process is not prevalent in this field of operations at the moment, although one airline located in Southeast Asia provided a training program to help an African nation develop its airline.

Regional Training Centers

The company would be glad to participate in regional training centers around the world in LDCs. Recently, a Mideastern country was willing to undertake such a project, but it has yet to get underway because each country approached is still insisting upon its own training facility on home soil. Hence the problem of attaining cooperation among sovereign nations on a regional basis is still very much unsolved.

Company C

Background

This company is a highly diversified manufacturer with sales in the billions per year. It is one of the nation's largest producers of many of its product lines. The subsequent discussion pertains to its chemicals and plastics operations. Its licensing procedures and relationships with developing countries differ by product line. This discussion is confined to situations where the company establishes joint ventures in developing countries, some wholly-owned and some on a shared ownership basis with the host countries. The company operates in about thirty-five developed and developing countries.

Licensing

All transfers of technology, whether with wholly- or partly-owned subsidiaries, take the form of license agreements. This is done to protect the company because of the highly uncertain technology policy in many developing countries. In addition, the Internal Revenue Service looks more favorably on the imputation of R&D expenditures to affiliates abroad under a licensing system than through a less structured relationship. The company generally establishes an agreement with a host country for a fixed term. After its expiration provisions are included for a cancellation by either party on appropriate notice. These agreements generally run for an initial period of ten years. If a shorter period is involved, it usually is at the insistence of the host government. The company believes it is difficult to convey the essence of its technology in as short a period as five years. Generally speaking, the fee structure is on a percentage of sales basis.

Patents are of peripheral value in this industry and principally permit management control over the licensee. The real value is in the technology itself. The patent serves to lessen competition and allow a marginally higher price. In other industries, however, the patent may be the sine qua non of the technology that is transferred.

Restrictive Business Practices

The antitrust problem is severe, and many American companies would be freer in licensing their technology if part of the consideration of the deal were for the recipient country to stay out of the American market. Some LDCs are insisting on their right to sell anywhere in the world. The reverse side is that American companies know that they will be unable to enforce a restriction on the licensee's marketing of a product regardless of what the original license may say. Consequently, some companies may decide that if it is impossible to restrict trade and technology, they will not get involved at all. The consequence of these decisions is a limitation on the amount of technology actually transferred.

MNC-LDC Relations

The company does not sell hardware. However, it is aware of the demands of many of the developing countries for only new equipment. These demands are associated with some unfavorable experiences of the LDCs with inflated prices charged for some inferior equipment. The company itself never has been involved in such a situation, and the experiences of most LDCs with most industries have been pervasively favorable. But it is obviously essential that the LDCs themselves distinguish among the different industries and companies with which they deal. It is important that all situations not be treated alike or in black and white terms. The company could not envisage any advantage in shipping inferior equipment in any event. Such equipment is likely to break down soon, and since the company is interested in evolving mutually beneficial longer-term relationships with the LDCs, such practices would simply be counterproductive. The only justification the company could advance for this behavior by certain multinationals might be the problems they may have had in repatriating currency from the LDC. In effect, overpricing the equipment might be one way of getting the money out of the LDC in the face of currency exchange limitations. In a more open environment there would be no need or justification for such behavior.

The company generally deals with those LDC companies of whom it has had some previous knowledge, either as a customer or in some other capacity. It generally does not deal with total strangers.

Know-How

The essence of technology transfer is contained in the know-how and not in the license per se. The investment that the company makes in putting together its whole operation and establishing a framework of supporting infrastructure is the basis out of which this know-how develops. The benefits of this transfer, including the technology, can have a very long, useful life.

Value of the Technology

The company tries to establish and maintain value for the technology to be transferred to both developed and developing country recipients. On occasion, the company has had some difficulty conveying its sense of the worth of this technology to the LDCs. The LDCs are sometimes unwilling to accept the value of the technology that has been demonstrated elsewhere in the marketplace. The company has found itself confronted with LDC arguments that the incremental costs of the transfer of the technology are all that the LDC should pay. In contrast, the company cited some earlier experiences with Japan, where it found a willingness on the part of both the Ministry of International Trade and Industry (MITI) and the Bank of Japan to recognize what the technology was worth and pay a fair price for it.

Competition

Considerable competition is involved in transferring technology. This competition cuts across the full range of activities, from building the plant, to purchasing materials to construct it, to making and selling the product. In pricing the total package the company identifies its own unique advantages as well as those of its competitors. If the company determines that its own position is inferior to the competition, it will withdraw.

Equity Interest

Where the company has an equity position, it is more likely to become involved in managing the entire project from the outset, including the construction of the plant. In certain other instances, it may merely design the project and allow the affiliate to undertake the actual construction. In effect the company provides the basic information required to design the plant and then it supplies the know-how to operate it. In the case of an affiliate, the company would probably supply intangible services such as marketing. However, if it is dealing with an independent, the company simply does not have the resources to provide such services to the LDC company.

When it feels there is a market to be served in the LDC, the company generally takes the initiative to transfer the technology or otherwise get involved. This is particularly true if the market is in the process of otherwise being foreclosed. If the company has no equity position, it is likely to be the more passive partner to the transaction, as might be the case where there is simply a license involved. An equity position, in fact, provides the opportunity as well as the compulsion for the company to make a larger contribution in order to assure the success of the endeavor.

Technology Transfer Versus Export

In many instances a relationship is worked out with an affiliate either to maintain or further penetrate the market in that country. While in some instances this technology transfer may not on its face be too appealing, the alternative is to lose the market completely. Competition is such that if the company does not get involved, another one will.

Vintage Technology

Great difficulties must be overcome if older technology is to be used again. There will be a lack of manpower that fully understands the process; in fact, the technology may almost be forgotten and in effect have to be reinvented. The information needed to rejuvenate some of these technologies is available, but what is lacking is an economic incentive to recreate this technology on the part of the people who possess knowledge of it. Industry might view quite cynically attempts by the government to recreate this knowledge through a subsidy or other means. In a practical sense, the size of the market that is likely to be generated for the vintage technology is likely to be quite limited. Some of the company's technology is simply not suited to the LDCs because of their market limitations.

If a technology is used that is less efficient, but more suited to the local resources, the LDC desire to enter the export market likely could not be satisfied, because the initial technology could not compete effectively for very long.

The company does not deal extensively in the transfer of products or machinery. However, some years ago, in response to a developing country request, it did dismantle a plant and reconstruct it, and it has been operating successfully ever since. The company has transferred some older technology to LDCs from time to time simply because the scope of the project and the investment involved did not support the expenditures associated with more modern equipment. The company was willing to work on this basis since it had an equity position. Economically this transfer has turned out successfully. Similarly, in an equity position, the company would not shave the technology transferred for environmental or safety reasons, even if the LDC recipient was willing to do so. The company adheres to its own standards, which may be more stringent than the local ones.

LDC Goals

The primary goal for a preindustrial society should be to achieve self-sufficiency in food. Once that is attained, industrialization can be attempted. The reason most LDCs seek to industrialize is to earn foreign exchange. Without an adequate agricultural base these efforts are foredoomed to failure.

There does seem to be a conflict between the LDC's goal of increasing employment and its desire to increase foreign exchange. Taiwan was cited as a country that successfully reconciled these conflicts. That country started in the early 1950s with very simple labor-intensive activities, and moved forward gradually. It also made certain of a solid agricultural base at the outset of its industrialization process. Korea seems to have proceeded successfully in a similar fashion.

Scale of Operation

The capital costs involved in increasing the capacity of a plant from ten to twenty million pounds of output a year to something like ten times that amount are not proportionate to the increase in the scale. Hence it is possible for an LDC to achieve economies of scale and operation and export the excess production in order to earn foreign exchange. The possibilities for scale economies are very great in this field of endeavor.

Host Country Environment

It is most important that the LDCs create a receptive environment to the transfer of technology. To accomplish this, the owner of the technology must be convinced he has some control over how it is used and where the product to which it contributes goes.

LDC Perceptions

It is important for the LDCs to realize that profitable operations do not occur instantaneously but take a long gestation period. There must be a realistic perception among these countries of what can and cannot be accomplished. Otherwise, any company in a position to transfer technology may very well choose to work with other developed countries rather than with developing ones, because of the limitations of its resources.

Training

The company prefers to develop training facilities within the LDC itself. However, it is important to make certain that the trainees are committed to perform on the job after training and not move elsewhere in the country. Otherwise one may end up as a trainer for the particular skills needed throughout the entire nation. Even some of the comparatively sophisticated countries do not have enough skilled people to go around.

Personal contacts are the best way to transfer technology.

Skill Level

Plant operators who actually run the unit are probably the most critical element in the operation of the whole project. One of the problems in the LDC labor market is the general upgrading of jobs because of the imbalance in education. For example, at home certain jobs must be filled by high school graduates, whereas in some LDCs the very same job occupants might have bachelors degrees in chemical engineering. There is overtraining and overeducation in some LDCs for certain selected skills.

Competitiveness

It is important to gear operations to the export market, even at the outset. If it is oriented only to the local market and conditions, the product or project is not going to stand up. This situation is similar to that created by countries establishing artificial barriers to protect their own industries. Over the long haul, these barriers will not sustain the industry.

Labor Intensiveness

Effort has been made in certain product lines to reduce the scale of operation to make it more compatible with the markets to be served. In terms of the flexibility of varying production techniques however, the manufacturing or production operation itself is not susceptible to any significant manipulation of the manpower requirements. Most of these opportunities occur in the peripheral areas. If one is not careful, he ends up, fairly quickly, with make-work operations rather than the substitution of labor for capital in an efficient manner. Such situations are artificial and will not withstand the test of time, especially if the LDC is seeking to make its product competitive in order to earn foreign exchange. Limitations are imposed on this transfer of labor for capital by safety requirements. If labor substitution is pushed very far, it is most likely to turn up in the support service area, such as an enlarged white collar bureaucracy in the firm's office operations.

R&D Facility

Because the costs associated with developing an R&D facility and its infrastructure are very large, it probably can only be supported in centralized facilities and in well-advanced areas such as the United States. It is difficult to transfer this capability even to some advanced developed countries, to say nothing of the less developed ones. The company itself has found it more economical to centralize R&D operations that cut across many divisions rather than to establish separate entities for each.

South-South Transfer

This process may have some attractiveness, particularly if it is to flow from the more to the less advanced of the group. In the former countries, the operating skills and background associated with some of the less advanced technology still exist, and it is possible for them to transfer the know-how to other developing countries more easily than from the United States.

Company D

Background

Among the largest in its field of operations, the company operates plants in more than forty developed and developing countries. Its net annual income has been around $175 million. The specific product line discussed involves agricultural products, chemicals, and pharmaceuticals. From the company's perspective, in most instances technology transfer goes from a parent company to a subsidiary which may be partially owned by the host country. In a few situations the company has a minority shareholder interest, though it prefers 100 percent ownership.

The transfer of technology from the company to the recipient country takes the form of an investment. The production process involves a basic manufacturing plant and a separate formulation or dosage plant. The former produces the primary raw material or basic chemical, and the latter provides further processing and packaging. A dosage plant can get its basic material from a plant located anywhere in the world. The company has both types of plants in the developing countries.

Overall Technology Transfer

Since the company is involved in actual plant construction in many of these countries, the transfer of technology starts from the moment the plant is designed. Local people have to be trained to dig the hole and pour the concrete, and to undertake the other tasks involved in constructing a factory. A high level of skill is required in the production of either the basic manufacturing material or the dosage plant. While the processes themselves have been highly mechanized, a considerable amount of judgment still is required in the production. A lot of local materials are put into the finished product and decisions must be made as to their quality.

Some of the machinery used in the production process is quite sophisticated, such as that used to package the product. This piece of equipment relies on locally produced cardboard and bottles, and sometimes problems occur when these items do not conform to the established tolerances for the packaging equipment. The company then has to negotiate with the local industries to get their products up to the standards. If these producers are still unable to make the adjustments, the company tries to import the material from abroad. However, this may require an import license, which the host country may refuse to issue on the grounds that it has the internal capacity to manufacture such products and that the company should make use of that capacity. This can lead to a basic conflict between the company and the host country. The company tries to anticipate fully all of these problems and spell them out in advance. However, given the complexity and uncertainties of the operation and the fact that the company cannot control the quality and availability of the auxiliary materials, problems continue to occur.

In many instances the host country appreciates the utility of the end-product that the company produces but fails to recognize the necessity of having available all the support and auxiliary industries essential for successful production.

The company looks to the indigenous market of the host country in deciding whether to invest. It also considers the political stability of that country and whether its market might

Technology Transfer in Practice

be closed to exports in the not too distant future. Once it concludes that the economics can be satisfied, it starts from the ground up to deal with all of the problems associated with constructing a plant. It has learned from experience to take nothing for granted and to investigate the availability of all resources, including water, power, utilities, and roads.

The company is especially concerned with maintaining a safe and sterile atmosphere for its pharmaceutical products.

Productivity

The productivity of certain dosage plants established in developing countries some time ago has increased, in some instances up to ten times the initial output. The basic source of this improvement is the continuous research in which the firm is involved. This has increased yields and brought about significant price and cost reductions. Dosage plants with an initial ability to handle capsules at a rate of between 500 to 1,000 per hour per machine have been able to increase the pace of the operation substantially via productivity gains.

New Technology

The company would be willing to install some of its newer equipment in the LDCs. However there is a general reluctance on the part of the LDCs to spend the necessary foreign exchange. They feel that present equipment is working well enough. Countries persist in resisting these changes, even though the net result might be improved earnings and increased foreign exchange. One explanation for this seemingly inconsistent behavior is the relative inexperience of many of the officials in some of these countries, who are not yet in a position to appreciate the potential advantages in upgrading and expansion.

Repair and Maintenance

Sometimes the maintenance of equipment proves troublesome when local sources are incapable of handling the problem in an efficient manner. The net result is that the operating plant in the LDC may develop its own workshop to maintain and repair equipment, which is not ordinarily done in an industrial country.

Brain Drain

There is a problem when individuals from LDCs choose not to return to their own countries after they have received training from abroad simply because no jobs are available there that would utilize their new skills.

Training

The company provides training to locals in particularly high skill or critical areas needed for the operation of the plant. These include plant managers, finance managers, marketing specialists, and those with similar functional responsibilities. This training may be provided in one of the company's plants located in another LDC and continues until such time as the trainee is presumed to be capable of operating on his own. He then returns to work in the company's plant located in his home country. Training is also provided in lower skill levels, including distribution and marketing of the product. At the outset, the company checks very carefully with other companies who have already built plants there to determine the quality of the local builders, architects, and others needed to construct a facility.

The company is confident that, given the time and opportunity, it could train the local people to operate a plant efficiently. Generally speaking, the facilities and perquisites of employment with Company D far exceed the quality of those available in locally-owned firms. The people themselves are generally highly motivated, willing to work, and responsive to the economic incentives of steady employment.

Workers in the local subsidiary company at the upper or medium skill levels may have considerable mobility. Their skills are transferable to alternate employment opportunities in their country or in others, including the United States. However, for such lower skilled operations as warehousing or production, there is a larger pool of labor; hence the mobility of workers in these categories is quite limited.

LDC Government

The process of establishing the investment involves dealing

with many government entities. The first thing that Company D must determine is the corporate form and financial structure of its subsidiary. The company also must investigate the availability of local financing once the plant is established. It also must secure an operating license, which may specify the type of plant it is going to build and the kind of product to be manufactured. In some cases these countries will restrict the quantity that can be produced in the plant and, even if demand for the product exceeds expectations, the plant will be held to these limitations. The LDC government does this to protect its local industry, whether it is owned privately or by the government.

Repatriation of Capital

The company tries to the fullest extent possible to get a guarantee in advance of its operations on the repatriation of capital and dividends. It also negotiates on the form of royalties which it may want on these products, as well as on its management fee. These royalties would be paid by the affiliate to the MNC. The current trend is for LDCs to disallow payment of such royalties and management fees. Wherever possible, under such circumstances, the company tries to compensate for these changes by adjusting its expected payout in the form of profits and dividends. However, regulations regarding these payments are also becoming stricter.

Some of the LDCs' complaints about MNCs' repatriating excessive amounts of capital may very well stem from misunderstandings about the company's capital structure and the relationship between debt and equity capital, as interpreted under the old British colonial system and under present conditions. Many LDCs interpret a company's capital structure as consisting only of its equity capital and not a combination of its debt and equity positions.

Developed-Developing Country Relations

Developing countries do not seem to appreciate the changed conditions of today as opposed to ten or fifteen years ago. In the earlier period, multinationals were anxious to invest in developing countries, but this is no longer the case. Companies are more discriminating and feel that, on balance, the risk-reward relationship might be more favorable toward investment in a developed country. The MNCs appreciate their social and moral responsibilities toward the developing countries and are prepared to do their part, but not at the same time that they are being subjected to inflamed political rhetoric and accusations of exploitation by the developing world. Many developing countries have placed so many restrictions on the operations of the MNCs that, given the opportunity, the companies would choose not to invest in them.

Because of the excessive restrictions and the deteriorating political environment, U.S. companies are becoming disenchanted about investing in such nations. This is probably true of companies in other industrial countries as well—including those with a reputation for social awareness and concern. Nevertheless, companies in these countries cannot repeatedly undertake investments that do not provide an adequate return. The LDCs must be aware of the contributions MNCs have made to their economies, not only the direct benefits associated with making the products themselves, but the training of the people in establishing the plants and the creation of ancillary industries to support the initial manufacturing project. Through the process of import substitution, considerable amounts of foreign exchange also have been saved.

Risks and Rewards

Because of the higher risks and uncertainties associated with investments in LDCs, it is normal to expect the return of an investment there to be larger than that for one in a developed country.

MNC and Local Government–Owned Firm Competition

The prices at which the company's subsidiary can market its products are often established by the government. The company knows that its own plant is very much more efficient than a government-owned plant, and that in some instances its production costs are substantially below those of the government operation.

The company views the technology process as encompassing the entire operation of the firm—from the very outset of plant construction to the marketing and distribution of the product. It considers its principal competition to be the very sophisticated European companies who are seeking a foothold in these markets.

The company's competitive advange over the LDC-owned company is in the production process, with its higher output for given levels of input. The company also has superior marketing and distribution skills to get the product to the ultimate consumer. In many instances, the government of an LDC may overestimate the potential market for a particular product and request the company to produce to that level. At such times, even though the company knows it can efficiently achieve the level required, it also warns the government that the market will not be able to absorb that volume. Should it proceed under such circumstances, the government will simply end up with a warehouse full of material as testimony to its lack of realism or sophistication.

Scale of Operation

In the United States the costs per unit of output are lower than they would be in foreign plants where the scale of operation is much smaller.

Regional R&D Centers

The prospects for these regional institutions are quite limited for a number of reasons. The skills required and the qualified people needed to man such facilities probably exist only in half a dozen or so industrial countries around the world and even in these countries the availability of some of the requisite skills is limited. For the lesser skill levels, while the concept sounds attractive, the practical implementation of such centers is very difficult to achieve because of political, geographic, and religious problems.

Priorities

The LDCs desperately need to establish their own priorities in a rational way. They must avoid setting goals for political or prestige purposes. Presumably we are past the era of large steel mills and local airlines as necessities for all countries around the world. It often would be more appropriate to establish plants that manufacture clothing or cheap tiles for housing. At all levels, there is a constant problem of transferring technology for both simple and sophisticated operations. Because of the commercial nature of these industrial operations, governments, whether of the United States or of the developing countries, simply lack the capacity to transfer this technology effectively. The major responsibilities belong with the companies involved and not with the governments.

One of the most important things that the LDCs can do in regard to technology and development generally is to establish priorities in terms of their own needs. They need an organized plan to tell them where they are going, where they would like to be, and how to get there.

Company E

Background

Company E is a highly diversified international corporation, active in both developed and developing countries, with net sales per year in excess of $5 billion. It has manufacturing facilities around the world.

MNC-LDC Relations

The company spokesman described a classic evolution of a situation in an LDC where the MNC operated a company that was then taken over by the government. The MNC continued to supply equipment to that operation from an industrial site, and over time, as the market within the LDC expanded, the MNC established manufacturing facilities within the LDC to service the first company. Technical support for this purpose was received from the home country facilities until such time as the new company was thoroughly integrated into the LDC economy. Gradually there has been a diminishment of the equity ownership until the home company now retains forty percent, while the LDC retains the balance. With this new company there is a continuing technology and licensing agreement. This enables the LDC company to keep current with the technological advances of the home company. The company generally deals with LDC government representatives.

Many LDCs simply cannot compete on a worldwide basis with an industrial country-based facility, even though their wage levels are very much lower, because their productivity is also lower. Another contributing factor is restrictions imposed by the LDC government on the firm's ability to hire and fire people.

The company deals with LDC companies in the private sector in which it has as much as 100 percent equity. It also deals with entirely government-owned facilities. In either case the whole process of transferring technology takes time and during this period the company has the opportunity of selling equipment at a profit to the overseas affiliate or licensee.

Licensing

Where the MNC has a majority interest in an affiliate, a general relations agreement will prevail. This agreement stipulates that each of the affiliates will pay a certain percentage of sales into a common technical pool. In return, that affiliate will have access to all of the developments that are made in any other facility within the multinational corporation complex. This provides for a free flow of information and visits to home company facilities. However, where the MNC holds a minority interest, the general relations agreement is replaced by a specific licensing agreement. It may stipulate rather precisely the products that are to be covered and licensed, rather than extending to the LDC affiliate the availability of anything within the MNC complex. The licenses generally run from five to ten years, with ten years the more usual.

The company occasionally sells technology but prefers not to, unless the potential buyer is insistent on that procedure. Most of the licensing procedures of the company are based on a percentage of sales.

The MNC tries to stipulate limitations on marketing arrangements at the outset of a licensing agreement. At the expiration of these agreements, however, the company's affiliates may try to expand into other markets or simply to go it alone. The affiliates do this primarily to earn foreign exchange, but given the high technology content of the product produced, they are unlikely to succeed over the long term.

The process of investing overseas is generally a very time consuming and drawn-out process of offer, counteroffer, and negotiation with the prospective host country. In terms of initiative for these investments, the MNC sometimes will make an unsolicited proposal to the government; conversely, an LDC government may go through a formal worldwide procedure of seeking competitive bids.

Renewals of initial licensing agreements are not automatic. It is possible for a recipient government to allow the arrangement to lapse and operate on its own for a while, but most likely, where there is a good relationship between the MNC and the host country, the country will seek to renew a licensing arrangement as new technological breakthroughs occur.

Adaptation to LDC Markets

The company has set up manufacturing facilities in many LDCs. In negotiating with individual countries, it attempts to adapt the scale and type of operation to the market that will be served. The MNC attempts to set up a facility that will utilize local labor at least in the assembly process. In some countries, the market simply will not justify the kind of capital expenditures that might be required to make all the parts and components of the finished product locally. In these cases, the company will ship in various parts for assembly in the LDC.

Differentiation among LDCs

The demand for technologically sophisticated final products varies among LDCs with the level of their own economic development. Thus the upper tier members of the LDC group, such as Mexico or Iran, would not want a manufacturing facility that did not embrace the highest level of technology. They are not interested in investing their money in facilities that may be classified as obsolete by western industrial standards. Other countries are sensitive to the problems of switching to newer and more efficient technology, not only because of the cost, but because the labor-saving newer procedures would have a detrimental effect on local employment. In one country that is very much aware of the need for creating employment, the government is grappling with this problem through the policy decisions it makes regarding urban or rural development. It has opted for the latter in the hope of keeping people from moving to urban centers and simply adding to the unemployment there.

Vintage Technology

The company is quite capable of providing any older technology that might be required. Where there is an LDC interest, the company may have available the capital equipment and other facilities of this older technology which it can make available at attractive prices to LDCs. At present, however, no such sales are taking place. Some LDCs recognize the appropriateness of selecting a less advanced technology that provides more than adequate service and that the people can maintain and operate.

Training

In sophisticated operations involving advanced technological equipment, supplying companies provide training programs to assist the locals to operate the facilities. These facilities are established in the home country, and the companies generally select and screen participants in these programs. This training requirement would be made explicit in the initial contract with the LDC. In some cases the candidates for the programs developed by the LDCs have been more than adequate, whereas in others the company has had a high rejection rate, primarily because of a lack of education and of exposure to working in the kind of environment that is required to operate a reasonably sophisticated piece of equipment.

R&D Capability

This factor varies greatly among countries in accordance with their level of development and other factors. Korea was cited as having great potential to develop its own R&D facilities, whereas some less wealthy countries in other cultural environments would be unable to do so within a reasonable period of time.

The company believes it is most successful in transferring technology when it can develop a subsidiary in a country and set up an operation, providing it with whatever specialized support is needed at the outset from an industrial country source, but establishing training programs and gradually making more use of indigenous labor and management.

The company has found, on a worldwide basis, including other industrial countries, that it is most efficient to bring its R&D and development capability into close proximity to its manufacturing facilities and plants.

The company transfers equipment abroad to LDCs from either its European or U.S. sources, depending upon which seems more appropriate. There is no firm company policy on this procedure.

Risk and Reward

There still seems to be some lag in the LDC's recognition of the full and true development cost of some of the finished products; hence, it would be very difficult to sell technology outright at what would be construed as a fair price. The company does try to establish appropriate estimates of desired rates of return related to the risks it anticipates in individual countries, though it does not have any fixed formula for determining such decisions. From the perspective of the LDC the price might be much too high, whereas from the point of view of the MNC, the asking price should reflect all of the costs associated with development of this product. Because of this, the company believes it can do better over time if it has an operating company in an LDC environment in which it has a substantial equity position. This procedure will enable the MNC to generate a sufficient flow of profits over time to achieve proper compensation.

Repatriation

In the face of increased restrictions by LDCs, it is becoming difficult to repatriate earnings and capital. The company works constantly to persuade government officials that it is in

Technology Transfer in Practice

their long-term interest to permit repatriation. Without it, the MNC would have no incentive to invest new technology or capital in the LDC.

Unbundling

In regard to the possibility of LDCs pushing for this kind of procedure, the MNC would probably accommodate itself to any kind of arrangement posed. However, by unbundling and separating the services from the hardware and operational aspects of the transfer, it would become more difficult to convince a prospective buyer of the value of the service by itself, due to its intangible nature. The company is aware of the difficulties of convincing any buyer in a developing country of the value of consulting services and the reason so much should be paid for them. The LDCs would have difficulty perceiving the value of such unbundled technology, even if its value were based upon a historical or reproduction cost basis. There is often difficulty in perceiving why a product that represents the embodiment of so much intellectual capital should cost so much. For these reasons Company E would prefer that this process not develop.

Nationalization

The company has experienced the nationalization of its operating facilities abroad. In some cases assets were simply expropriated; in others, proper compensation was made.

Indigenous Populace

In establishing a new subsidiary, the company generally provides supervisory and other high skills for some time in order to train the local population from the bottom up. In the phaseout of foreign personnel, the last to go is generally the managing director of the operation. The company tries to have a local person in a position of full managerial responsibility by the time this changeover occurs.

Reverse Flow

As an example of the reverse flow of technology the MNC cited an Asian country that took a process developed in an industrial country and, in a matter of a few weeks, digested the system and in fact improved its effectiveness by about 1000 percent. This accomplishment so impressed the manager of the industrial country facility that he took the innovation back home to incorporate into his own procedures. The time to test a particular process was reduced from one hour to something between thirty seconds and one minute. This of course is an exception to the general flow of technology, but it does occur and with substantial benefits.

Job Creation

In negotiating with the LDC, the MNC generally tries to stipulate the number of jobs for locals that are expected to result from a particular project. This is the opposite of a demand by the LDC that a specific number of jobs must be created. In one instance, the company badly overstated the amount of employment it could create over a ten-year period because it overestimated the size of the market in this particular country. This estimate was not affected by the cost side of the operation because the cost of labor was a very small part, but even that, in fact, turned out to be too costly for the market. The size of the market constrained the size of the investment so that it was impossible to create a situation where the company could economically employ 150 people.

Negotiations

The MNC has found that the LDCs will hire outside consultants to handle the technical aspects of the negotiation. In some instances these consultants come from other LDCs that are more advanced, or from some international organizations. The LDCs are perceptive enough to recognize the need for technical guidance in this area.

Competition

The extent of competition varies from country to country. Where a traditional relationship between the MNC and the LDC has been established, this relationship is likely to be extended in a more or less informal manner. Other countries prefer to set up a competitive situation each time. Compared

Technology Transfer in Practice

with twenty years ago, it is clear that the competition today is much more intense.

LDC Planning

In anticipation of the U.N. conference, this spokesman suggested some government-to-government planning assistance to help the LDCs establish their own priorities within a particular industrial sector and perhaps for the entire economy. The U.S. government could provide money and maybe even the talent, in the form of outside consultants, to the LDC government for such purposes.

Company F

Background

This company is involved in industrial, automotive, and materials handling activity. It works with LDCs primarily through investments, though on occasion it has sold or leased technology through licensing. It prefers 100 percent ownership in its foreign affiliates, but as this is often prohibited in certain countries, it will accept less than 100 percent and even a minority ownership if required in order to gain access to an important market and if there is a reliable local partner available. Since this firm is primarily a supplier of intermediate goods or components used to produce a final consumer product, it tends to follow its major customers in their location decisions. Its reason for investing in foreign countries is generally to follow one of its customers, taking into account both the potential in that particular market and relations with the customer worldwide. The alternative would be to cede this market to the company's competitors, since both transportation and tariff costs usually make servicing the customer from an outside production base inefficient.

Generally speaking, the smaller scale of the operations required in an LDC makes them noncompetitive on a worldwide scale. In addition, the high border protection in many LDCs makes it very difficult, in the absence of tariff remissions or other forms of government subsidies on exports, to reexport this material on a competitive basis.

While the company formerly looked for desirable acquisitions, the customer-dependent nature of the business often meant that there were no desirable existing facilities. Thus the company's more recent expansions have tended to include new plants. This obviously accelerates the growth of countries that already have developed larger and more sophisticated markets.

Capital-Labor Costs

The lower wages in many LDCs usually are more than offset by lower labor productivity. Also, since this industry is capital-intensive, wages are a small portion of total costs. The biggest cost burden, then, is in the capital area. Import tariffs that the firm must pay also represent a major cost element. The relative costs of production in developing countries are artificially impacted by their governments, which tend to push the price of labor up as well as restrain the cost of capital.

Comparative LDC–Industrial Country Operations

A number of differences were noted between a plant in an LDC and one producing a similar product in the United States or another industrial country that has a large domestic market as well as the ability to service markets worldwide. First, the production run was much shorter in the LDC plant because of its high product diversity; that is, it had to manufacture not only product A but products B, C, and D as well—all from the same production line. In an industrial country there would be separate lines for each product.

Second, much more capital was invested in the larger plants in the industrial countries. Third, the company has developed some specialized tooling to service the LDC market. One such change includes the development of various inserts needed in the smaller, more diversified LDC production lines. This task requires a reasonably high level of skill and the local labor force has usually proved trainable to handle it. Finally, the high tariff levels imposed on the imports of certain types of tooling equipment prompted the company to adjust its production processes in the LDC to extend the life of the tooling itself. For example, while slowing down the feed and speed of the tooling equipment may reduce its efficiency somewhat, these slow-

Technology Transfer in Practice

downs effectively double its productive life. This trade-off is made necessary by the comparative prices of capital and labor in the LDC. Less capital and more labor are therefore used in these countries than in the United States.

The critical element in determining the size of the plant in an LDC turns on the amount of the tariff the country chooses to impose on its imports. For this company, direct production labor makes up about five to ten percent of total costs, while total labor comprises about 15 percent. There was only a limited opportunity to substitute labor for capital in the peripheral areas, such as materials handling, primarily because of the weight and size of the materials being handled and the items produced.

Export Versus Investment

Given the risks, uncertainties, and administrative problems, Company F would prefer to export to LDC markets rather than to establish a plant in these countries. However, the LDCs' desire to industrialize and to increase the labor content and value added aspects of their exports are primary factors that have forced the company into making these investment decisions.

From the company's perspective the LDCs' attitude toward foreign investment could be summarized as follows (and often an LDC's ambassador or industry minister seeks out Company F and tries to sell it on this basis): We are offering you a market, we want you to bring in your technology, for which we will pay you a royalty; we will also pay you a return on your own capital investment. However, we will not let you earn any money on our own resources, such as debt capital, that you use within our country. In this way, among others, LDCs try to limit the amount of earnings the MNC can generate and repatriate to the size of the capital it imports.

In many of the countries where Company F operates, repatriation of capital is permitted up to about 12 percent of that invested, with a 30-percent withholding tax. These rates can vary somewhat, but this is the general limitation. Some of these limitations are indexed to allow for inflation; but where this procedure is not followed, there can be a major problem.

The lack of regional technology centers in these countries has not provided any obstacle to investment.

Indigenous Labor Force and Training

This company's investments tend to be in the more advanced developing countries of the world, where the local labor force generally is capable, flexible, and responsible, although sometimes extensive training programs are required. The initial trained cadre trains others, who in turn pass on their skills. These programs have succeeded to such a point that a cadre of workers from one South American country is now used for training purposes on a worldwide basis. Other than for initial training, the company tends to minimize its use of U.S. or other industrial country personnel in the LDCs in favor of the fullest possible use of indigenous populations. The local labor force usually has functional capability, so that once trained the workers are able to do their jobs. The literacy rate is not high in these countries. It was noted that the American management attitude is quite different from any other in the world—it is much more pragmatic and features a willingness on the part of management to mix in the operational side of the business, which still is not the case in many other countries.

Vintage Equipment and Scale

The company will generally use smaller and older equipment to service smaller markets, whether they are located in developed or developing countries. This older and used equipment is quite efficient and the product quality is equal to that of the newer machines. This performance level is essential, since sometimes the products of all the company's machinery are sold on a worldwide basis and therefore must be interchangeable. The older machinery is quite capable of producing today's product. However, some LDCs forbid the use of any but the newest equipment and insist on machines with greater capabilities than the product blueprint actually requires. These types of restrictions are self-defeating for the LDCs, since they may result in higher capital costs that make the resulting product less competitive on a worldwide basis.

Company to Company

In making an investment decision the company would work very closely with its private-sector counterparts in the LDCs, including operating people, lawyers, and trade associations. They would explore such matters as the political stability of the country as well as the economic viability of the project itself. They would undertake careful research on the availability and local supply of needed resources, as well as the market potential for the product itself. The selection of a local business partner is a major part of the decision process. In many LDCs, the company would build into its own operations the repair and maintenance facilities that outside suppliers would normally provide in the industrial countries. In dealing with private-sector groups in these developing countries, the company is much more comfortable with industrialists, that is, those who understand production problems and processes and have realistic expectations about the nature of the particular industry. Industrialists also better understand that profits are a long-term objective rather than something to be gleaned on a quick trading basis. The businessmen in the LDCs are quite competent and hardheaded about the realities of the business and do not share the rhetoric of the political figures.

South to South

The company thought that Brazil, among other countries, was quite likely to have its own technological base in a few years, from which it would be able to export technology to other LDCs.

Unbundling

One needs to classify the LDCs into two groups: those that want foreign investment and technology for balance of payments reasons; and a second, richer group that may want only the technology. The company would be willing to provide consultative services through a management contract in lieu of an investment in certain cases where there is an incentive to do so. Technology does not really flow through licensing and patents, per se, so that a much more extensive involvement in the full range of management decisions must be maintained in order to transfer the technology effectively. Pure licensing transactions may be more attractive to smaller U.S. companies who may view the technology itself as the product they are selling. In many LDCs the company now finds itself, in effect, selling its technology, since the licenses are for a limited period—say 5 years—and are nonrenewable. The company really does not like this policy but resorts to it when it has no alternatives, or if the particular technology itself is of limited value. A joint venture with a foreign counterpart gives the company additional leverage for dealing with that company and seeing that the technology is properly used. It also assures a larger return from the investment and technology. When only a license is employed or the technology is unbundled, the company may, in effect, be creating its own future competition. Since this particular industry does not change very rapidly, it must be very careful about practices that simply license the technology developed by the company at its own expense to these potential competitors.

Competition Among MNCs

Because the competition among MNCs seeking foreign investments is quite intense, LDCs seem to be enjoying a fairly comfortable market position. The competition with other MNCs generally takes the form of the size of the plant to be built, the number of jobs to be created, and the level of exports promised.

Impact of LDC Inflation

Another obstacle to the investment and technology transfer process is the soaring inflation in Latin America and many other LDCs. In some cases, repatriation limitations are indexed to take account of inflation. Even so, the devaluations of local currencies which reflect that inflation reduce the effective returns from the operation, making it less attractive. A company must consider the write-down of overseas assets now required under Financial Accounting Standards Board Statement No. 8 for the purpose of U.S. balance sheets. This process tends to depress the value of the parent company shares.

Moreover, the amount of after-tax earnings now required just to stay even with inflation and take account of replacement costs is growing. These "decapitalizing" factors mean that what may superficially appear to be excessive profits in the host country may in fact be an inadequate return to warrant the technology-related investment.

Long-Term LDC Objectives

The most appropriate goal for the LDCs in their efforts to achieve economic development is to maximize the number of jobs that can be produced on an efficient basis using a worldwide scale of operations as the measuring stick. This means the basic orientation of the country should be toward a blend of imports and exports. The country should not pursue merely an import-substitution approach, which is often based on the "black-box" myth regarding what technology can do. The company has found that many of the rules requiring local content, number of jobs, etc., are quite arbitrary, too generalized, and really not geared to the needs of individual industries. They are therefore counterproductive insofar as the LDCs are concerned. (Ministry officials, it may be noted, often do not understand the economic or technical aspects of business; although, of course, their frame of reference is different. For example, even if the country could benefit economically from rationalizing component production by using output from a plant in a neighboring country, government officials might perceive the latter as a rival and be unwilling to allow any "interdependence.")

In summary, the company felt that it was most appropriate for the LDCs to concentrate on their own traditional comparative advantages and to pursue policies of free trade. This approach would redound to their benefit as well as to that of the industrial countries, including the United States.

Appendix A:
Agenda, United Nations Conference on Science and Technology for Development

I. Science and Technology for Development:

   A. The choice and transfer of technology for development;
   B. Elimination of obstacles to the better utilization of knowledge and capabilities in science and technology for development of all countries, particularly for their use in developing countries;
   C. Methods of integrating science and technology in economic and social developments;
   D. New science and technology for overcoming obstacles to development.

II. Institutional Arrangements and New Forms of International Cooperation in the Application of Science and Technology:

   A. Building up and expanding institutional systems in developing countries for science and technology;
   B. Research and development in the industrialized countries in problems of importance to developing countries;
   C. Mechanisms for exchange of scientific and technological information on experiences significant to development;
   D. Strengthening of international cooperation among all countries and the design of concrete new forms of international cooperation in the field of science and technology for development;

E. Promotion of cooperation among developing countries and role of developed countries in such cooperation.

III. **Utilization of the Existing United Nations System and Other International Organizations:**

Utilization of the existing United Nations system and other international organizations to implement the above goals in a coordinated and integrated manner.

Appendix B: Annotated Agenda, United Nations Conference on Science and Technology for Development

The specific guidelines to be followed in approaching subitems of agenda items I, II, and III are shown below.

I. **Science and Technology for Development:**

 A. The choice and transfer of technology for development;
 B. Elimination of obstacles to the better utilization of knowledge and capabilities in science and technology for the development of all countries, particularly for their use in developing countries.

 Subitems A and B should be considered individually and together in terms of the following points, including the linkages between national development plans and programmes and international technological relations, as well as the factors which create:

 1. The state of technological dependency and analysis of the factors which increase or decrease such dependency and of the various degrees of technological dependency. Analysis of the difficulties encountered in the processes of transfer and selection of technology, and of the determining factors in the transfer of

Extracted from the U.N. "Guidelines for the Preparation of National Papers," A/CONF.81/INF.1.

technological capacity and the importation of technology. The analysis should take into consideration the need to strengthen the capabilities of developing countries to choose and adapt technologies in accordance with their national policies and priorities, particularly considering various relevant factors, for example, the practices of transnational corporations, technological monopolies, the barriers to the flow of advanced and proprietary technology, limited technological infrastructures and so on;

2. Assessment of national measures taken in the context of A and B, particularly measures taken to rationalize imports of capital goods, to promote scientific and technological information systems, to develop extension services capabilities on the part of research institutes, consulting firms and technology development enterprises, including those necessary for the adoption of integrated national policies for technology transfer and development and those necessary for coordinating the evaluation and negotiation of technologies;

3. From the country's experience, analysis is to be presented with regard to the following (and other) obstacles which have impeded the formation and/or attainment of the country's intention to apply science and technology to development:

   a. Lack of appreciation of the role of science and technology in development;
   b. Lack or inadequacy of a scientific and technological infrastructure;
   c. Lack of access to scientific and technological information;
   d. Inadequate contact between endogenous research and development and technology users;
   e. Inadequate or unsuitable education and training;
   f. Emigration of scientific and technical manpower ("brain drain");
   g. Lack or inadequacy of planning;
   h. Lack of adequate criteria for the choice of tech-

nologies that are appropriate to the development objectives of the country;
  i. Shortage of *entrepreneurs* and managerial skills;
  j. Unsuitable national or international institutional systems for science and technology;
  k. Insufficient financing resources (domestic or foreign exchange) for investment;

4. The formulation of appropriate recommendations to solve the stated problems through actions at the national, regional, interregional, or global level.

C. Methods of integrating science and technology in economic and social development;
D. New science and technology for overcoming obstacles to development.

Subitems C and D should be considered individually and together, putting emphasis on the following:

1. A detailed analysis of the present state of technological capability, the application of technology to all sectors of the economy, in particular to the production sectors, and science and technology policy as an integral part of the over-all national planning process;
2. An analysis of the national measures adopted and envisaged by each country in order to:

  a. Enhance the capabilities of technological supply from both internal and foreign origin;
  b. Promote the application of science and technology for rural development;
  c. Stimulate the demand for local scientific and technological output (technology plus personnel) within all sectors of the national economy, so as to make optimum use of local scientific and technological capacity;
  d. Foster the role of basic science, applied science, engineering, social science, experimental develop-

ment, and technological services and the balance between the resources devoted to them;
e. Foster the role of extension services;
f. Foster approaches to overcoming economic, social and environmental problems created by newly introduced technologies;
g. Promote the interaction between the scientific and technological systems and other systems, particularly the sectors of production;
h. Popularize science and technology with emphasis on bringing about a change in attitudes towards the use of science and technology in the development process;

3. A discussion on new science and technology for promoting development with specific examples of new and longer-range scientific and technological developments, which if properly applied, hold promise for development;
4. Recommendations to facilitate the short-range and long-range solutions of concrete problems as they are detected, paying particular attention, *inter alia,* to:

a. Those measures directed at ensuring a faster substitution of foreign technologies by those that may be generated by local scientific and technological capacity;
b. Mechanisms for the control and selection of technology;
c. Mechanisms to regulate and canalize foreign investment as devices for the transfer of technology;
d. Measures to facilitate the unpackaging of technology;
e. Measures to regulate industrial property.

II. **Institutional Arrangements and New Forms of International Cooperation in the Application of Science and Technology:**

A. The building up and expansion of institutional systems in developing countries for science and technology.

Subitem A should be considered individually, placing emphasis on:

1. The national conceptualization of the scientific and technological systems of each country;
2. The diagnosis of the current national situation with regard to the specific subitem involved. Assessment of the scientific and technological infrastructure capacity as typified by economic and social research areas;
3. The measures adopted by each country to solve the problems thus described;
4. The role played by international cooperation in the solution of problems faced by the external sector in the developing countries' economies that limit development of national systems of science and technology, describing possible actions that should be taken in the short term, medium term and long term in order to use to the maximum the benefits of such cooperation;
5. The elaboration of measures taken to ensure the optimal use of human resources; promotion of the training and continued improvement of the technical experts needed for the development of the national scientific and technological system; and formulation of policies directed at curbing the exodus from the developing countries of trained personnel.

B. Research and development in the industrialized countries in regard to problems of importance to developing countries.

Subitem B should be considered individually, placing emphasis on the following:

1. The national papers prepared by industrialized countries should include a description of the current status of their respective scientific and technological potentials with particular emphasis on quantitative data wherever feasible about:

a. The direction given to such potential in the context of national socio-economic development objectives;
b. The identification and wherever possible the classification of those scientific and technological activities of benefit to the developing countries;
c. The trends in levels and kinds of resources applicable in various ways to solving national, regional and world-wide problems, and particularly those of developing countries;
d. The distribution of such potential by economic sectors;
e. Investments in scientific and technological activities applicable to development problems in relation to and/or in terms of gross domestic product;

2. As a result of this general description, measures taken by each developed country to facilitate access of developing countries to the research and development programmes that are relevant to the solution of their development problems should be listed; new measures to improve the existing situation should also be specified;
3. Special reference should be made to the role that international cooperation could play in enhancing the participation of developing countries in the scientific and technological development efforts carried out in industrialized countries, including the role in this respect of international financial cooperation;
4. Developed countries should analyze the relative success or failure of their respective policies of international development cooperation as they affect efforts of developing countries to build endogenous science and technology capabilities;
5. In the context of their national papers the developing countries should submit comments on this subitem. The comments should provide analytical explanations on encouraging research and development in and by industrialized countries to be oriented in new, more

Appendix B

effective and practical ways towards the solution of concrete development problems in the developing countries.

C. Mechanisms for the exchange of scientific and technological information and experiences significant to development;
D. The strengthening of international cooperation among all countries and the design of concrete new forms of international cooperation in the fields of science and technology for development;
E. The promotion of cooperation among developing countries and the role of developed countries in such cooperation.

In the discussion of the above subitems, the following should be achieved:

1. A general description of current mechanisms for exchanging scientific and technological information on a national, regional, interregional and world-wide basis should be given;
2. A general review should be prepared of current technical, scientific and technological cooperation schemes on a subregional, regional, interregional and world-wide basis;
3. As a result of the above descriptions, a diagnosis from a national perspective of the effectiveness of such schemes as tools to strengthen and develop technological capabilities in developing countries should be prepared;
4. Special reference should be made to the role that cooperation among the developing countries could play through the establishment of joint action schemes that make possible:

    a. The establishment of joint programmes in the field of scientific and technological activity to solve specific problems of three or more countries.

b. The introduction, and the joint use, of the established infrastructure with a view to making maximum use of them;
c. The organization of the exchange of information and experience, particularly as regards the scientific and technological capacity of each country;
d. The organization of systematic information programmes;
e. The formation of systematic training programmes for specialized personnel;
f. The strengthening of the negotiating capacity of developing countries regarding the acquisition of technology, including the designing of a joint negotiating model;

5. National points of view on the role of developed countries in support of collaborative programmes and projects among developing countries should be defined, including those activities regarding the encouragement of imports of technology from developing countries, financial cooperation for the technological development programmes arising from the cooperation schemes among developing countries; training programmes for scientific and technical personnel in developing countries; and access to their scientific and technological information systems;
6. Recommendations should also be made concerning the ways and means to strengthen international cooperation among all countries, especially between developed and developing countries, including if appropriate proposals for new schemes and mechanisms.

III. **Utilization of the Existing United Nations System and Other International Organizations:**

Utilization of the existing United Nations system and other international organizations to implement the objectives set

Appendix B

out in Economic and Social Council resolution 2028 (LXI), paragraph 3, section I, in a coordinated and integrated manner.

Recommendations should also include measures that should be adopted to strengthen coordination, increase efficiency of existing mechanisms, or establish new action mechanisms, or to restructure international organizations in the field of scientific and technological cooperation for the benefit of all countries and, in particular, developing countries.

Appendix C: Agenda and Participants, International Economic Studies Institute Workshop, June 22, 1978

Agenda

Welcome and Introduction, *Dr. Timothy W. Stanley*
Workshop Overview, *Dr. Samuel M. Rosenblatt*
Appropriate Technology: Obstacles and Opportunities, *Professor Gustav Ranis*
Demand for Appropriate Technology, *Professor Nathaniel H. Leff*
Luncheon Speaker, *The Honorable Jean Wilkowski*
Technology and Employment Constraints on Optimal Performance, *Professor Howard Pack*
Technology Transfer in Practice, *Dr. Samuel M. Rosenblatt*
Discussion of Issues Affecting the United Nations Conference on Science and Technology for Development
Summary, *Dr. Timothy W. Stanley*

List of Participants

Dr. Timothy W. Stanley, President and Trustee, IESI
Dr. Samuel M. Rosenblatt, Consultant to IESI
Dr. Gusav Ranis, Economic Growth Center, Yale University
Dr. Nathaniel Leff, Graduate School of Business, Columbia University
Dr. Howard Pack, Swarthmore College
The Honorable Jean Wilkowski, Coordinator, U.N. Conference on Science and Technology for Development, Department of State

(The above are listed in the order in which they spoke at the meeting.)

Appendix C

Dr. *Jack Baranson*, President, Developing World Industry and Technology, Inc.
Mr. *John Borland*, Economist, IESI
Mr. *Karl L. Buschmann*, Consultant to IESI
Mr. *Harlan Cleveland*, Director, Program in International Affairs, Aspen Institute for Humanistic Studies, and Trustee, IESI
Mr. *Ronald L. Danielian*, Vice President and Trustee, IESI
Mr. *William Eilers*, Office of Science and Technology, Agency for International Development
Mr. *Curtis Farrar*, Policy Planning Staff, Department of State
Ms. *Anna Fotias*, Subcommittee on Science, Technology and Space, Senate Commerce Committee
Mr. *Robert W. Grim*, Assistant to the President, International Division, Owens-Illinois, Inc., and Trustee, IESI
Dr. *Perfecto Guerrero*, Science Attaché, Embassy of the Philippines
Dr. *Jean M. Johnson*, Operational Studies and Analysis International Division, National Science Foundation
Mr. *William Kawood*, Auditor, General Accounting Office
Mr. *Frank Kinnerley*, Department of State, OES Bureau, Office of Advanced and Applied Technology Affairs
Dr. *John L. McLucas*, President, COMSAT General Corporation, Trustee, IESI
Dr. *José Pastore*, São Paulo University, São Paulo, Brazil
Ms. *Carolyn Rhodes*, Legislative Assistant to The Honorable Clarence D. Long
Mr. *Ludwig Rudell*, Department of State (T/CST)
Dr. *Wilson E. Schmidt*, Professor of Economics, Virginia Polytechnic Institute and State University, and Trustee, IESI
Dr. *Albert Small*, Office of Science and Technology, Department of Commerce
Professor *Jack Taylor*, Chairman and Professor of Economics and Management, St. John Fisher College, and Senior Lecturer in Economics, University of Rochester
Ms. *N. Ethelyn Thompson*, Secretary, IESI

Professor Lawrence White, Graduate School of Business Administration, New York University

Mr. Thomas W. Wilson, Jr., Aspen Institute for Humanistic Studies

Mr. Christopher Wright, Office of Technology Assessment, U.S. Congress

Appendix D: Senior Advisory (Steering) Committee for Institute Research Project on Technology and the World Political Economy

Dr. Jack N. Behrman, Professor of International Business, University of North Carolina

Dr. Hylan B. Lyon, Jr., Manager, Advanced Planning, Texas Instruments, Inc.

Professor Stephen Magee, Finance Department, University of Texas

Professor Edwin Mansfield, Department of Economics, University of Pennsylvania

Dr. Gustav Ranis, Professor of Economics, Economic Growth Center, Yale University

Seymour Rubin, Esquire, Executive Director, American Society of International Law

Dr. Charles Wolf, Jr., Head, Economics Department, The Rand Corporation

Ex officio

Dr. Timothy W. Stanley, President, International Economic Studies Institute

Dr. Samuel M. Rosenblatt, Senior Consultant and Technology Project Director, International Economic Studies Institute

Appendix E: About the International Economics Studies Institute

The International Economic Studies Institute is a nonprofit, publicly supported, tax-exempt research and educational organization, established under the laws of the District of Columbia. It is nonpolitical and dedicated to objective research on international economic issues of concern to Americans.

Institute publications reflect the views of their authors and not necessarily those of the trustees or officers. The institute's responsibility is to present them to the public as useful contributions to understanding and policy choices.

The institute has depository arrangements with libraries of the Federal Reserve Board in Washington; the Federal Reserve Banks of Chicago, Dallas, and Minneapolis; Harvard and Princeton Universities; the University of Hawaii; The Brookings Institution; the Council on Foreign Relations; the Brooklyn Business Library; and the New York Public Library.

Trustees

David C. Acheson, partner, Drinker, Biddle & Reath (formerly partner, Jones, Day, Reavis & Pogue, senior vice president and general counsel, Communications Satellite Corporation, former special assistant to the secretary of the U.S. Treasury, and U.S. attorney for the District of Columbia)

Harlan Cleveland, director, Program in International Affairs, Aspen Institute for Humanistic Studies (formerly president, University of Hawaii, and former U.S. ambassador

Appendix E

to NATO, and assistant secretary of state for international organization affairs)

Ronald L. Danielian, executive vice president and treasurer, International Economic Policy Association, and director, Center for Multinational Studies (formerly director, Office of Research and Analysis, U.S. Travel Service, U.S. Department of Commerce)

Isaiah Frank, William L. Clayton professor of international economics at the Johns Hopkins School of Advanced International Studies (formerly executive director of the Commission on International Trade and Investment Policy)

Robert W. Grim, assistant to the president, Owens-Illinois International

Jacob J. Kaplan, author and consultant in international finance and economics (formerly U.S. representative to the managing board of the European Payments Union, assistant coordinator for foreign assistance, U.S. Department of State, and consultant to the National Commission on Materials Policy)

Walter L. Lingle, business consultant; member of IEPA board of directors (formerly executive vice president, Proctor & Gamble Company, and former deputy administrator of the Agency for International Development)

John L. McLucas, president, COMSAT General Corporation (formerly administrator, Federal Aviation Administration, former under secretary and secretary of the air force, president of Mitre Corporation and assistant secretary-general of NATO for scientific affairs)

Wilson E. Schmidt, head of the Department of Economics, Virginia Polytechnic Institute and State University (formerly deputy assistant secretary of the U.S. Treasury for research)

Timothy W. Stanley, president, International Economic Policy Association (formerly visiting professor, School of Advanced International Studies and former defense advisor, United States Mission to NATO)

Officers

President and Treasurer, *Timothy W. Stanley*
Vice-President, *Ronald L. Danielian*
Secretary and Assistant Treasurer, *N. Ethelyn Thompson*

Appendix F: Institute Publications

Books

Raw Materials and Foreign Policy (IESI, 1976; Westview Press, 1977).

Technology and Economic Development: A Realistic Perspective edited by Samuel Rosenblatt, with a foreword by Harlan Cleveland (Westview Press, 1979).

Contemporary Issues Series

No. 1 *The Raw Material and Commodity Controversy* by Dr. Harald B. Malmgren, Fellow, Woodrow Wilson International Center for Scholars (October 1975).

No. 2 *International Economic Dimensions of Secondary Materials Recovery* by Professor Ingo Walter, New York University Graduate School of Business Administration (December 1975).

No. 3 *Dependence and Vulnerability: The United States and Imported Raw Materials* by Jacob J. Kaplan and Timothy W. Stanley (December 1977).

No. 4 *Recent International and Raw Materials Developments: A U.S. Perspective* by John Borland (in progress).

Appendix F

No. 5 *East-West Technology Transfer and U.S. Export Controls: A Structural Analysis* by Samuel M. Rosenblatt and Karl L. Buschmann (in progress).

Additional papers will be forthcoming.

Other Publications

Basic Data for Estimating U.S. Mineral Requirements and Availability: A Research Compendium (February 1976).

Conference Papers

"The United States, the European Community, and Raw Materials," Center for European Studies Graduate School, City University of New York, May 1975.

"Strategic Resources and the North-South Dialogue," National Security Affairs Conference, "Toward Cooperation, Stability and Balance," June 1977.

"Contingency Planning for Material Resources: A Contemporary Perspective of Resource Problems," 1978 Engineering Foundation Conference, December 1977.

"International Problems and Raw Materials Supply," American Chemical Society, National Symposium on Critical and Strategic Materials, *Proceedings*, June 1978.

"The Security of Energy Supply: Some Preliminary Thoughts," a discussion paper for the Aspen Institute Committee on Energy, Aspen Institute for Humanistic Studies, July 1978.